TECHNICAL REPORT

National Guard Special Forces

Enhancing the Contributions of Reserve Component Army Special Operations Forces

John E. Peters • Brian Shannon • Matthew E. Boyer

Prepared for the United States Army

RAND ARROYO CENTER

The research described in this report was sponsored by the United States Army under Contract No. W74V8H-06-C-0001.

Library of Congress Cataloging-in-Publication Data

Peters, John E.
  National Guard Special Forces : enhancing the contributions of reserve component Army Special Operations
Forces / John E. Peters, Brian Shannon, Matthew E. Boyer.
    p. cm.
  Includes bibliographical references.
  ISBN 978-0-8330-6012-9 (pbk. : alk. paper)
  1. United States. Army Reserve—Organization. 2. Special forces (Military science)—United States. I. Shannon,
Brian. II. Boyer, Matthew E. III. Title.

  UA42.P48 2012
  356'.16—dc23

                                                                                          2012035743

The RAND Corporation is a nonprofit institution that helps improve policy and decisionmaking through research and analysis. RAND's publications do not necessarily reflect the opinions of its research clients and sponsors.

RAND® is a registered trademark.

Published 2012 by the RAND Corporation
1776 Main Street, P.O. Box 2138, Santa Monica, CA 90407-2138
1200 South Hayes Street, Arlington, VA 22202-5050
4570 Fifth Avenue, Suite 600, Pittsburgh, PA 15213-2665
RAND URL: http://www.rand.org/
To order RAND documents or to obtain additional information, contact
Distribution Services: Telephone: (310) 451-7002;
Fax: (310) 451-6915; Email: order@rand.org

Preface

This technical report presents research undertaken as part of a project entitled "Enhancing the Contributions of Reserve Component Army Special Operations Forces." The project was designed to explore ways to enhance the contributions of U.S Army National Guard (ARNG) Special Forces to ongoing U.S. military operations and to provide recommendations that might lead to "purpose-driven" ARNG Special Forces: that is, forces organized and employed to take advantage of civilian skills, language proficiency, or other attributes found predominantly within the ARNG Special Forces.

This report examines the options for recasting ARNG Special Forces in a "purpose-driven" way. It is intended for a readership centered within the U.S. Army Special Operations Command and its subordinate U.S. Army Special Forces Command. The report should be of interest to the ARNG Special Forces community and the 18 state adjutants general who oversee ARNG Special Forces within their states.

This research was sponsored by Lieutenant General John F. Mulholland, Jr., Commanding General, U.S. Army Special Operations Command (USASOC) and conducted within RAND Arroyo Center's Strategy, Doctrine, and Resources Program. RAND Arroyo Center, part of the RAND Corporation, is a federally funded research and development center sponsored by the United States Army.

The Product Unique Identification Code (PUIC) for the project that produced this document is USAS105647.

The project points of contact are Brian Shannon, 703-413 1100, extension 5270; email bshannon@rand.org; and John E. (Jed) Peters, 310-393-0411, extension 6188; email jpeters@ rand.org.

Contents

Figures

Tables

Summary

This study was undertaken to help the commanding general, U.S. Army Special Operations Command (USASOC), develop options for enhancing the contributions of U.S Army National Guard (ARNG) Special Forces by making them a "purpose-driven" force rather than simply a copy of the active component (AC) forces under his command. The intent was to provide data to inform deliberations regarding the potential future direction of ARNG Special Forces.

Key Tasks

First, we examined and described the policy framework for ARNG Special Forces. This framework establishes the latitude available to USASOC in creating a purpose-driven ARNG Special Forces.

Second, the research sought to determine the strengths and weaknesses of ARNG Special Forces Groups. This effort involved conducting surveys, interviews, and a literature review to support comparison of AC and National Guard attributes, capabilities, capacities, and authorities. This comparison is the basis for subsequent recommendations, based on a consensus between the AC and ARNG Special Forces on those missions and tasks for which the National Guard units are well suited.

Third, the research developed a menu to offer USASOC as recommendations for developing options to enhance ARNG Special Forces contributions to USASOC. These options emphasized missions (e.g., Afghanistan village security) and units (e.g., ODA and ODB) that lie within the AC/ARNG consensus on the most suitable employment of ARNG Special Forces.

Lines of Inquiry

In executing the key tasks in the research design, we pursued three lines of inquiry. First, we examined large amounts of data provided by USASOC and the ARNG. These data included:

- Records of hazardous duty pay, which assisted in tracking individual combat deployments.
- Unit deployment records, which established when specific ARNG Special Forces units were deployed, e.g., operational detachments Alpha (ODA), operational detachments Bravo (ODB), advanced operational bases (AOB), and special operations task forces (SOTFs).
- Training and qualification records, which indicated how many personnel were qualified in their primary military occupational specialty (MOS), and how many personnel pos-

sessed additional skills, e.g., military free fall parachuting, sniper, scuba, and advanced special operations training (ASOT).

These data provide a basis for comparison with the AC Special Forces in terms of frequency of deployment, numbers and types of units deployed, and individual deployment histories.

Second, we conducted an online survey that asked questions about the background, prior service, qualifications and deployments of the respondents, and their views on the relative merits of ARNG Special Forces.[1] The responses to the survey helped us determine the civilian skills resident among Guardsmen respondents, their tolerance for future, additional deployments, and other factors that might be revealing of potential niches or pockets of unique, high-value contributions for ARNG Special Forces employment in future operations.

Finally, we conducted policy-level interviews in which we queried the adjutants general for the principal states hosting ARNG Special Forces units, assistant secretary of defense–level officials with reserve affairs responsibilities, and others down to and including the senior leadership of the ARNG Special Forces Groups themselves. These interviews established the participants' views of ARNG Special Forces capabilities and limitations, and their ideas for enhancing contributions from these units.

What the Research Found

- USASOC had hoped to find civilian police skills, analytical skills, and language skills that could serve as the basis for unique ARNG unit contributions within Special Forces. Although many valuable civilian skills are present within the ARNG, they do not exist in the densities that would enable the ARNG to build specific units around them. They do constitute a rich pool of individuals who might be potential volunteers for individual mobilization, but even on an individual basis, they cannot be involuntarily mobilized under current authorities because of their civilian skills.

- Language experience in the ARNG Special Forces is wide, but of limited depth. It seems doubtful the National Guard could be a significant source for language-qualified Special Forces personnel without considerable, additional effort.

- There is an important interdependence between the ARNG Special Forces competence and the AC Special Forces units' confidence in the National Guard that rests on deployments. The ARNG Special Forces must deploy at some reasonable frequency in order to be competent and to be trusted by their AC Special Forces counterparts. Insufficient deployments put the ARNG Special Forces on a spiral to irrelevance.

- There is a conditional consensus emerging between both the AC and ARNG about their niches. Those subscribing to this consensus generally believe the ARNG ODAs, ODBs, and individual augmentees are suitable for theater security cooperation activities, unconventional warfare, and foreign internal defense, perhaps with an emphasis on the "non-kinetic" aspects of the latter two. Individual augmentees can serve useful functions within SOTFs, joint SOTFs, and AOBs, according to this view.

- There is much that USASOC could do to enhance the utility of the ARNG Special Forces. Closer, more frequent coordination and greater commitment to predictability

---

[1]   The survey and a detailed analysis of respondents' replies appears in Appendix C.

and lead time would be important next steps. Plan aggressively with the ARNG Special Forces on a time line that they can manage to overcome their DMOSQ shortfall. Finally, renew directed training affiliations and mission letters and hold more coordination conferences so that all the ARNG Special Forces have visibility into the Playbook (the USASOC planning calendar that reflects the units identified for future deployment: their assignments, the time frame, and similar details), their future deployments, and the mission-essential tasks they must master in order to be prepared.

The study's ultimate recommendations appear below in Figures S.1 and S.2. The colored numbers to the left of each listing in Figure S.1 suggest an order of implementation, based upon a logic reflecting the authorities available to USASOC and the costs of implementing each option. Figure S.2 illustrates. According to Figure S.2, USASOC should implement those actions whose costs are low, and that can be done on the organization's own authority.

These include employing ARNG Special Forces for tasks including theater security cooperation (TSC), joint combined exchange training (JCET), foreign internal defense (FID), unconventional warfare (UW), Combined Joint Task Force–Horn of Africa Building Partner Capacity operations, and extended training operations. They also include employing ARNG Special Forces to ease the operations tempo (OPTEMPO) for the AC units. These recommendations also consider the appropriate units for employment, emphasizing ODA, ODB, and SOTFs as the most appropriate size formations for ARNG Special Forces to command and control. When USASOC seeks to tap individual skills, it could operate an Internet website to solicit volunteers based upon their civilian skills. Finally, the inexpensive, unilateral recommendations advocate for the renewed use of mission letters to specify mission-essential tasks for each ODA, and to ensure that all ARNG Special Forces undergo some minimum number of operational deployments to maintain their skills and the confidence of their AC counterparts, with whom they typically operate when deployed.

Figure S.1
Recommendations

 • Employ ARNG Special Forces for recommended tasks (TSC, JCET, FID, UW, Horn of Africa–like, Afghan village security, etc.)
- Deploy to manage active component OPTEMPO
- Emphasize employment of ODA, ODB, and SOTF
- Operate Internet site to solicit volunteers based on their civilian skills
- Renew use of mission letters
- Guaranteed deployments to maintain skills

 • Regular Army advisors at SF company level

③ • More Special Forces Qualification Course quotas and support
- Extended Playbook
- Revitalize directed training alignment (DTA) relationships
- Sponsor more coordination and planning conferences
- Sponsor nominative assignments for promising senior ARNG Special Forces officers

 • Seek authority for access to ARNG Special Forces for non-named operations
- Create mobilization sites at DTA active component home station
- Create proportionate force structure to facilitate rotations

RAND *TR1199-S.1*

Figure S.2
Implementing the Study's Recommendations

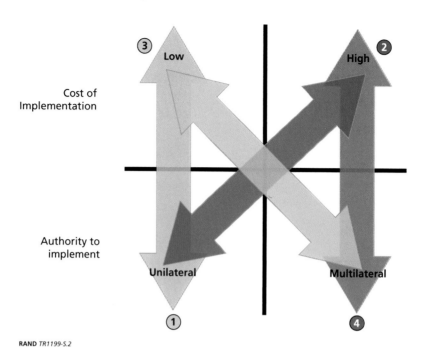

RAND *TR1199-S.2*

The second class of recommendations—those that are unilateral but expensive—contains a single recommendation. USASOC should return to the practice of assigning Regular Army advisors at Special Forces company level. Virtually everyone we encountered had a positive view of this practice and saw it as a very effective way to transmit recent operational experience and tactics, techniques, and procedures into the ARNG.

The third category of recommendations includes those actions that are relatively inexpensive but require multilateral agreement and coordination. There are five such actions. The first of these is for USASOC to send the ARNG more Special Forces Qualification Course (SFQC) quotas and to task U.S. Army Special Forces Command (USASFC) to work with the ARNG units and state adjutants general to prepare Guardsmen candidates, support them and their families during the course, and produce a higher graduation rate. The second recommendation in this category is for USASOC to extend the Playbook and share its contents earlier so that ARNG units have better insight into when they will be mobilized next, where they are likely to be deployed, and what missions they are likely to perform. Third, USASOC should revitalize directed training alignments (DTAs) between the AC and ARNG Special Forces. Ideally, the ARNG mobilization sites should be co-located with their DTA AC unit and they should deploy together. Falling short of that, the Regular Army company-level advisors should come from the DTA AC unit, and the DTA units should coordinate all collective training with the ARNG units aligned with them. Fourth, USASOC and USASFC should sponsor more conferences to conduct planning and coordination with the ARNG units. For example, force generation conferences and the process of building the Playbook should involve the ARNG Group commanders. Finally, in order to enhance senior leader (lieutenant colonel and above) capabilities within the ARNG Special Forces, USASOC/USASFC should sponsor nominative assignments that would afford promising ARNG officers the opportunities to deploy in AC staff and command jobs and to gain experience under the direct supervision of AC seasoned experts.

The final category of recommendations—both expensive and requiring multilateral coordination and agreement—includes three actions. USASOC should seek authority to access its ARNG Special Forces involuntarily for non-named operations.[2] Such authority would make it much easier to employ the ARNG to manage AC operations tempo. Second, USASOC/USASFC should create mobilization sites at the DTA home stations so that the ARNG Special Forces would mobilize and fall in on their AC counterparts. Third, insofar as USASOC must sustain a smooth rotation of forces in overseas contingency operations and direct interchangeability of units is desirable, USASOC should create proportionate units in the ARNG. Finally, USASOC in cooperation with the National Guard Bureau might offer financial incentives for active duty Special Forces separating from the Army to join an ARNG Special Forces unit. This would help increase DMOSQ ratings.

---

[2]   At the time of this report, there were broader efforts ongoing by the Department of Defense to review and potentially change the legal hurdles with involuntarily mobilizing reserve component units for non-named operations.

Acknowledgments

We are indebted to many people for their assistance with this study, first and foremost, the adjutants general or their designated representatives from Alabama, Colorado, Florida, Mississippi, Rhode Island, Texas, Utah, and West Virginia. We thank the members of the 19th and 20th Special Forces Group who met with us or corresponded with us over the course of the study. In the Office of the Secretary of Defense, our thanks go to the Assistant Secretary of Defense for Reserve Affairs and the Deputy Assistant Secretary of Defense for Reserve Affairs (Readiness). At the National Guard Bureau, we thank the acting Director, Army National Guard, and his staff for their assistance. We also benefited from interviews, telephone calls, and email exchanges with senior AC Special Forces personnel familiar with the National Guard, and we are grateful for their insights. Finally, at USASOC, our thanks to LTG John Mulholland, who sponsored this study, and to his staff, including Mr. Larry Deel and Mr. George J. (Jim) Lane III for their direct support of the project, arranging calendar time, facilitating meetings and briefings, and generally helping the project team. We also benefited from the support of our action officer at U.S. Army Special Forces Command, Major Israel Villarreal, to whom we are also indebted.

At RAND, we thank our colleagues Fred Wassenaar and Matt Boyer, respectively, for assistance with the web-based survey and the review of relevant law and policy governing the employment of the National Guard. We thank our reviewers, Susan L. Marquis, James T. Quinlivan, and Robert G. Spulak, Jr., for their helpful comments on an earlier draft of this report. Finally, we are indebted to our program director, Lauri Rohn, and her associate program director, Thomas Szayna, for their advice, support, and guidance throughout the course of the project.

Acronyms

AC	Active Component
AOR	Area of Responsibility
ARFORGEN	Army Force Generation (process)
ARNG	U.S. Army National Guard
ARSOF	Army Special Operations Forces
ASI	Additional Skill Identifier
ASOT	Advanced Special Operations Training
BOG	Boots on the Ground
CDR	Commander
CJSOTF	Combined Joint Special Operations Task Force
COCOM	Combatant Command
CS	Combat Support
CSS	Combat Service Support
DMOSQ	Duty Military Occupational Specialty Qualification (typically the percentage of personnel in a unit qualified in their duty military occupational specialty)
DoD	Department of Defense
DoDI	Department of Defense Instruction
DTA	Directed Training Alignment
FID	Foreign Internal Defense
FORSCOM	Forces Command
FTS	Full-Time Servicemember
GSC	General Support Company
HHC	Headquarters and Headquarters Company
IMA	Individual Mobilization Augmentee
JCET	Joint Combined Exchange Training
JSOTF	Joint Special Operations Task Force

MACV-SOG	Military Assistance Command, Vietnam–Studies and Observations Group
MFP/MFP-11	Major Force Program/Major Force Program 11 (which supports special operations forces)
MTO&E	Modification Table of Organization and Equipment
MOS	Military Occupational Specialty
MRAP	Mine Resistant, Ambush Protected (vehicle)
NCO	Noncommissioned Officer
NGB	National Guard Bureau
ODA	Operational Detachment Alpha
ODB	Operational Detachment Bravo
ODT	Overseas Deployment Training
OSD	Office of the Secretary of Defense
POM	Program Objective Memorandum
RC	Reserve Component
REFRAD	Released from Federal Active Duty
SF	Special Forces
SFG	Special Forces Group
SOCOM	Special Operations Command
SOF	Special Operations Forces
SOTF	Special Operations Task Force
TAG	The Adjutant General
TSC	Theater Security Cooperation
USAR	U.S. Army Reserve
USARV	U.S. Army, Vietnam
USASFC	U.S. Army Special Forces Command
USASOC	U.S. Army Special Operations Command
USSOCOM	U.S. Special Operations Command
UW	Unconventional Warfare

CHAPTER ONE

Introduction

The U.S. Army National Guard (ARNG) contains two Special Forces Groups (SFGs): the 19th and the 20th. These two groups complement the active component (AC) groups: the 1st, 3rd, 5th, 7th, and 10th SFGs. Mathematically, the two National Guard SFGs constitute 29 percent of the total and represent a valuable asset, especially after nine years of war—and one of increasing value if they can be more appropriately organized, trained, equipped, and employed. And years of war have honed the capabilities of both the AC and the National Guard. Nevertheless, according to both AC and ARNG officials, using the National Guard is not as easy as it might be. The mobilization process is arduous, the time and resources available to support pre-mobilization preparations are finite, and command and control of ARNG units spread over 18 states is challenging. That said, access to ARNG Special Forces is not impossible. The challenge, rather, is to create policies and practices that will give them the lead time and predictability they need to prepare for deployment, and that will focus their deployments on missions, operations, tasks, and activities where they enjoy a comparative advantage while limiting the exposure of their vulnerabilities: risk management activities all competent commanders practice in tasking their units.

The commanding general, U.S. Army Special Operations Command, LTG John F. Mulholland, Jr., sponsored the research that produced this report. Broadly speaking, he sought options for enhancing the contributions of ARNG Special Forces.[1] As he described it, this effort should identify niches in which the National Guard part of his force could excel, and which might take advantage of their strong suits: skills from their civilian careers, language capabilities, perhaps depth of experience in other domains. He was also insistent that the project should identify options for making ARNG Special Forces a "purpose-driven" force rather than simply a copy of the AC forces under his command. In this regard, being purpose-driven meant being organized and employed to take advantage of civilian skills and attributes unique to the ARNG. General Mulholland offered anecdotes about ARNG ODAs [operational detachments Alpha] manned with police who were very effective in site exploitation, forensics, and similar skills that allowed them to root out the enemy in ways that AC units could not. He hoped that this study would reveal other areas of endeavor where the ARNG might have valuable expertise that could be brought to future fights.

The research sought to identify key factors shaping the options for ARNG Special Forces:

---

[1] The question of how to better integrate ARNG Special Forces is not new. For example, see LTC Wayne J. Morgan, *Reserve Component Special Forces Integration and Employment Models for the Operational Continuum*, Carlisle Barracks, PA: Army War College, April 15, 1991.

- The legal, policy, and regulatory forces that shape ARNG and Special Forces and options for employing them.
- The demand for Special Forces generated by ongoing combat operations, security cooperation activities, and established joint training requirements (e.g., joint combined exchange training).
- The supply of ARNG Special Forces and the factors that limit and constrain it (e.g., the frequency of recent involuntary deployments, numbers of members who are qualified as Special Forces soldiers, and so on).
- Finally, the skills, knowledge, and abilities that ARNG Special Forces personnel might bring to future deployments from their civilian lives.

Research Design and Lines of Inquiry

The project pursued three tasks and three lines of inquiry.

Key Tasks

First, we examined and described the policy framework for ARNG Special Forces. In this effort we reviewed the statutory, Department of Defense (DoD), Office of the Secretary of Defense (OSD), and service policies in order to summarize current policy, identify constraints and limitations on the employment of ARNG Special Forces, and trace and assess the history of ARNG Special Forces evolution.

Second, the research sought to determine the strengths and weaknesses of ARNG Special Forces Groups. This effort involved conducting surveys, interviews, and a literature review to support comparison of AC and ARNG attributes, capabilities, capacities, and authorities. The research outlined factors affecting ARNG Special Forces readiness and suitability for various roles and tasks.

Third, the research developed a menu of options to offer U.S. Army Special Operations Command (USASOC) as recommendations for developing alternatives to enhance ARNG Special Forces contributions to USASOC. This process included identifying niches and gaps where ARNG Special Forces might usefully be employed, and identified policy issues and constraints from task one that, if resolved, could produce ARNG Special Forces of greater utility than those available today.

Lines of Inquiry

In executing the key tasks in the research design, we pursued three lines of inquiry. First, we examined large amounts of data provided by USASOC and the ARNG. These data included:

- Records of hazardous duty pay, which assisted in tracking individual combat deployments.
- Unit deployment records, which established when specific ARNG Special Forces units were deployed (e.g., operational detachments Alpha (ODAs), operational detachments Bravo (ODBs), advanced operational bases (AOBs), and special operations task forces (SOTFs).
- Training and qualification records, which indicated how many personnel were qualified in their primary military occupational specialty (MOS), and how many personnel pos-

sessed additional skills (e.g., military free fall parachuting, sniper, scuba, and advanced special operations training (ASOT)).

These data provide a basis for comparison of ARNG Special Forces with the AC Special Forces in terms of frequency of deployment, numbers and types of units deployed, and individual deployment histories.

Second, we conducted a web-based survey that asked questions about the background, prior service, qualifications and deployments of the respondents, and their views on the relative merits of ARNG Special Forces.[2] The responses to the survey helped us determine the civilian skills resident among Guardsmen respondents, their tolerance for future, additional deployments, and other factors that might be revealing of potential niches or "sweet spots" for ARNG Special Forces employment in future operations.

Finally, we conducted policy-level interviews in which we queried the adjutants general for the principal states hosting ARNG Special Forces units, assistant secretary of defense–level officials with reserve affairs responsibilities, and others down to and including the senior leadership of the ARNG Special Forces Groups themselves. We also interviewed AC officers who were knowledgeable about ARNG Special Forces performance in recent operations either because they had been in the chain of command for those units or because they served on a staff at a headquarters that oversaw some dimension of ARNG Special Forces employment. These interviews informed our sense of relative capability, limitations, and constraints on the use of ARNG Special Forces, and also contributed insights suggesting possible niches.

Organization of This Report

The remainder of this report contains three chapters and three appendixes. Chapter Two presents the basic policy framework that shapes ARNG Special Forces, the supply of those forces, the demand for Special Forces generally, and the skills, knowledge, and abilities that reside within the ARNG Special Forces. The chapter concludes with a brief description of the interaction between the USASOC force generation process and the ARNG unit life cycle of alert, mobilization, deployment, and release from federal active duty.

Chapter Three presents our understanding of ARNG strengths and limitations. The chapter integrates material from the Special Forces survey with expert views gathered during the policy-level interviews and other discussions with senior Special Forces officers who have served in the chain of command or on staffs with recognizance over ARNG Special Forces. Based upon this understanding of ARNG Special Forces, Chapter Three concludes by identifying potential niches for ARNG Special Forces: operations and activities we believe are consistent with their strengths that take full advantage of their civilian skills and experience.

Chapter Four contains a menu of options that USASOC might pursue in order to enhance the performance and therefore the contributions of ARNG Special Forces. The menu in Chapter Four treats a broad set of issues that represent a variety of constraints on ARNG Special Forces performance, and suggests the remedies appropriate for reducing these constraints.

---

[2] The survey and a detailed analysis of respondents' replies appear in Appendix C.

Appendix A provides a brief history of reserve component SFGs. To those unfamiliar with Special Forces or reserve component Special Forces, the history provides background on how two SFGs exist in the ARNG.

Appendix B presents a redaction of the laws and policies that shape ARNG Special Forces and, in some instances, place constraints on them: access to them for use in non-named operations, the frequency with which they can be deployed, and similar considerations.

Appendix C contains the Special Forces survey questions, sampling strategy, and responses that supported the study.

CHAPTER TWO

ARNG Special Forces and USASOC

The relationship between the ARNG Special Forces and USASOC is a guarded one. The U.S. Army National Guard believes that the legal and policy environment contains requirements, obligations, and authorities that collectively entitle it to certain guarantees for force structure and equipment comparable to that of the AC, which USASOC ignores. USASOC staffers regularly note the difficulties in getting access to the ARNG and the cumbersomeness of their procedures. There is a tension between demand for Special Forces to sustain the current combat operations, and the ARNG's ability to generate fully qualified teams. And there is mutual suspicion; the AC Special Forces share stories of ineptness and incompetence on the part of their ARNG partners when deployed, and the ARNG Special Forces members share stories of grievance and improper treatment when deployed beneath an AC-dominated chain of command.[1]

This chapter begins by acquainting readers with the basic parameters of the legal and policy environment that shapes AC and ARNG Special Forces relations, the Special Forces supply and demand dynamic and the ARNG Special Forces role in satisfying that demand, and the skills, knowledge, and abilities of the ARNG Special Forces community. The latter half of the chapter describes the basic relationship between USASOC and the ARNG Special Forces, emphasizing the processes that involve the ARNG Special Forces with USASOC in determining what units will deploy, getting them trained, mobilized, and deployed, and then returned to the authority of their states.

There are four basic considerations that bear on ARNG Special Forces:

1. The legal and policy environment that shapes them and their use.
2. The supply of ARNG Special Forces and the qualities of that supply relative to the AC Special Forces.
3. The skills, knowledge, and abilities (SKA) resident within the ARNG Special Forces.
4. The demand for Special Forces generally arising from operational requirements from the combatant commands (COCOMs) and from USASOC.

---

[1] These characterizations result from hours of discussions with staff officers within USASOC and U.S. Army Special Forces Command (USASFC) and from conversations with ARNG Special Forces personnel over the period October 2010 through January 2011.

Legal and Policy Environment

This section highlights eight subjects that we believe have significant impact on ARNG Special Forces and their availability to support USASOC and ongoing U.S. military operations around the world. Appendix B provides a more comprehensive treatment of the legal and policy environment and its effects on ARNG Special Forces.

Department of Defense Instruction 1235.10

This Department of Defense Instruction (DoDI) is important in several regards. First, Enclosure 2 establishes that reserve component members should be notified up to 24 months in advance that they are being considered for mobilization, and indicates that mobilization orders should be issued as soon as it is feasible to do so. The DoDI indicates that the DoD standard for mobilization approval to mobilization date is 90 days, with a goal of 180 days. Adherence to this standard would do much to address the predictability that many of the ARNG personnel we spoke with indicated is essential for them to successfully integrate their National Guard obligations with the other aspects of their lives.

Second, DoDI 1235.10 directs officials to "ensure early consideration is given to the practical use of alternate workforce sourcing solutions (AC, DoD civilians, coalition forces, host-nation support, civilian contracted labor, technological solutions, other government agencies, nongovernmental organizations, private volunteer organizations, or other means available)." Former senior OSD officials indicate that no one understands this passage as a serious constraint on access to the ARNG Special Forces. If this is the case, the policy might usefully be redrafted to specify the level of cooperation and integration that DoD expects between its AC and ARNG Special Forces, especially given today's persistent conflict and the potential value that the ARNG brings as an operational partner.

Section 104b, Title 32, U.S. Code

The thrust of this section is that "the organization of the Army National Guard and the composition of its units shall be the same as those prescribed for the Army, subject, in time of peace, to such general exceptions as the Secretary of the Army may authorize . . ." Many of the officials we interviewed, especially state adjutants general and senior ARNG officials, claim that the Army has failed to meet the intent of this statute in not updating the ARNG Special Forces Groups' MTO&Es to comport with changes made to the AC groups. Specifically, these officials note that the ARNG groups have general support companies while the AC groups have general support battalions, the AC groups have special troops battalions that the ARNG groups do not, and the AC groups have four-company battalions while the ARNG groups have three-company battalions. They state that these organizational differences interfere with one-for-one interchangeability and the smooth rotation of units through the deployment cycle.

Other officials, including those in the office of the Assistant Secretary of Defense for Reserve Affairs, did not share this concern about mirror-imaging the AC units. They seemed satisfied that the AC and ARNG shared the same basic building blocks of ODAs organized into companies, which in turn are organized into battalions, and battalions that constitute the SFG. As a practical matter, it is difficult to set a specific set of criteria against which to determine that a National Guard unit's "composition shall be the same as those prescribed for the Army," because there are differences in the MTO&Es of the AC groups, especially in terms of vehicles.

Moreover, historically, Special Forces (SF) have demonstrated organizational flexibility to meet the needs of specific missions. This has been the case with the Son Tay raid in 1970, which employed a purpose-built force.[2] It was the case with the creation of the 46th Special Forces Company in Thailand during the Vietnam War era,[3] and in Vietnam itself, where the various reconnaissance projects (GAMMA, SIGMA, DELTA), MACV-SOG, and USARV Special Missions Advisory Group each were purpose-driven in their organizations.[4] These specialized organizations then became reabsorbed into SF after the missions ended.

Nevertheless, if the expectation is that ARNG units must be able to replace AC units on a one-for-one basis, then the ARNG units should have same-sized building blocks. That is, if the AC deploys its supporting units (sometimes referred to as "enablers") in company and battalion-sized packages, the enablers in the ARNG should be packaged in same-sized units.

Section 104c, Title 32, U.S. Code

This section notes, "To secure a force the units of which when combined will form complete higher tactical units, the President may designate [National Guard] units to be maintained in each State. However, no change in the branch, organization, or allotment of a unit located entirely within a State may be made without the approval of its governor." Thus, USASOC would need the support of the governors where ARNG Special Forces units are posted in order to undertake any major reorganization.

As a practical matter, we understand that state adjutants general have been able to "horse trade" units to achieve mutually desirable redistributions of forces. The recent movement of two ARNG Special Forces companies to Texas is a case in point.

State Missions

Domestic missions for ARNG Special Forces units vary from state to state. In at least one state, the lone SF company does not have any specific mission, although, like any ARNG unit, it could be activated by the state. On the other hand, a number of states integrate and utilize the SF units for a variety of missions, including counter-narcotics, search and rescue, disaster response, and command and control. In some of these cases, ARNG Special Forces personnel act in various capacities as law enforcement officers. Ultimately, they have proven important when called up by the state. States that do use ARNG Special Forces units in state emergencies include Florida, West Virginia, and Alabama.

The low level of usage for state missions is generally due to two factors. First, in many states the actual members of the SF unit do not reside in the same state as the unit. Quickly activating the unit is difficult. Secondly, some states have a large number of general purpose National Guard units capable of providing the needed manpower.

The current state missions pose minimal constraints for USASOC as it looks to develop a future plan for ARNG Special Forces. What risk and disturbances that would occur to a state by a change in ARNG Special Forces can be mitigated by helping transition other ARNG units in the state into the current responsibilities. This could be done for many of the missions the ARNG Special Forces currently undertakes. More importantly, one state discussed

---

2   Benjamin F. Schemmer, *The Raid,* New York: Harper & Row, 1976.

3   See the 46th Special Forces Company Association home page at http://www.46thsfca.org/.

4   Shelby Stanton, *Vietnam Order of Battle*, Mechanicsburg, PA: Stackpole Books, 2003.

retraining other ARNG assets in boat operations before an ARNG Special Forces unit moved to another state, because the capability had proven important in past emergencies.

Chain of Command

ARNG Special Forces are distributed across 18 states. In some instances, such as Rhode Island, the state has a single Special Forces company. In other cases, such as Texas, the state maintains two companies, but each is subordinated to a different battalion located out of state. The chains of command that result are often complex and, prior to mobilization, do not extend across state lines. Commanders do not rate their subordinate commanders in circumstances where their subordinates are located in another state. Thus the chain of command reflects limitations that are not found in AC units. Figures 2.1 and 2.2 present unit diagrams annotated to show their locations. As Figure 2.1 indicates, the 1-19th SFG has forces in five locations in three states. In the case of its A Company, its subordinate units are in three principal cities. The second battalion likewise has its forces dispersed over five cities in three states. The 5-19th SFG has similar circumstances, with its subordinate units in five cities over three states.

The story for the 20th SFG is similar to that of the 19th SFG. The 1-20th SFG draws its forces from six cities in three states. The 2-20th SFG likewise has its subordinate units in six cities across three states. The 3-20th SFG has its subordinate units in four cities in two states. Figure 2.3 illustrates the wide geographic distribution of ARNG Special Forces units, especially the subordination of groups and battalions to multiple states when not mobilized for federal service.

Similarly, the individuals who make up any one unit may actually live outside of the state whose National Guard he or she is a member of. Figure 2.4 indicates the locations where members of 20th SFG reside and shows that it has members residing in 44 of the 50 states. Soldiers must travel to their unit to conduct training.

These circumstances manifest themselves in three critical areas: administrative support, logistical support, and mobilization support. In the AC, units can expect assistance from their higher headquarters, which typically include staff officers and NCOs with expertise in administrative and logistics areas. This expectation is not fully realized in the ARNG, because the

Figure 2.1
19th Special Forces Group Unit Locations

Figure 2.2
20th Special Forces Group Unit Locations

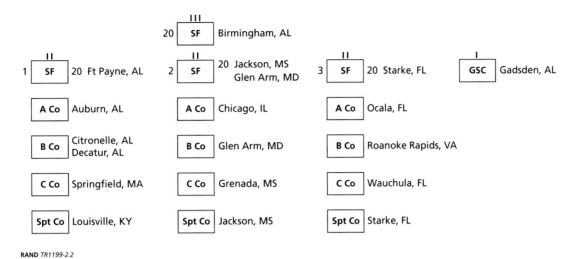

Figure 2.3
Geographical Distribution of ARNG Special Forces Units

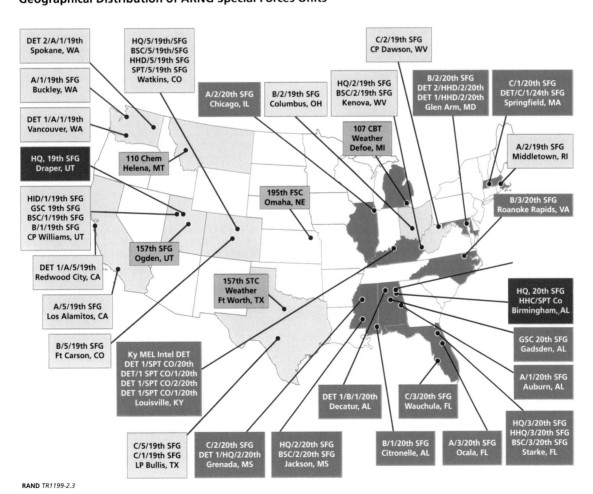

Figure 2.4
20th SFGs Members' States of Residence

Areas where the 20th SFC(A) does not have soldiers/families: Guam, Idaho, North Dakota, Puerto Rico, Rhode Island, South Dakota, Utah, Vermont, Virgin Islands, and Wyoming.

SOURCE: SIOPERS, March 2009.

RAND *TR1199-2.4*

1	Alabama	476	25.48%
2	Alaska	1	0.05%
3	Arizona	3	0.16%
4	Arkansas	7	0.37%
5	California	4	0.21%
6	Colorado	1	0.05%
7	Connecticut	6	0.32%
8	DC	9	0.48%
9	Delaware	4	0.21%
10	Florida	325	17.40%
11	Georgia	79	4.23%
12	Hawaii	1	0.05%
13	Illinois	98	5.25%
14	Indiana	14	0.75%
15	Iowa	3	0.16%
16	Kansas	5	0.27%
17	Kentucky	78	4.18%
18	Louisiana	21	1.12%
19	Maine	4	0.21%
20	Maryland	62	3.32%
21	Massachusetts	56	3.00%
22	Michigan	6	0.32%
23	Minnesota	2	0.11%
24	Mississippi	176	9.42%
25	Missouri	7	0.37%
26	Montana	1	0.05%
27	Nebraska	1	0.05%
28	Nevada	1	0.05%
29	New Hampshire	8	0.43%
30	New Jersey	15	0.80%
31	New Mexico	2	0.11%
32	New York	27	1.45%
33	North Carolina	106	5.67%
34	Ohio	6	0.32%
35	Oklahoma	5	0.27%
36	Oregon	3	0.16%
37	Pennsylvania	27	1.45%
38	South Carolina	20	1.07%
39	Tennessee	81	4.34%
40	Texas	28	1.50%
41	Virginia	69	3.69%
42	Washington	3	0.16%
43	West Virginia	1	0.05%
44	Wisconsin	16	0.86%
		1,868	100.00%

unit chain of command lacks authority prior to mobilization. Moreover, an in-state, alternative chain of command subordinates ARNG Special Forces to other state National Guard force structure for administrative purposes. As a result, especially among outlying units, they face their administrative, logistical, and mobilization issues largely on their own. Nevertheless, according to some ARNG Special Forces soldiers with whom we spoke, the resulting chains of command can generate significant workloads for the unit's full-time personnel (typically no more than four in a company of 87) without helping the unit substantially with its administrative, logistical, or mobilization-related issues.

In some instances, these arrangements have prompted feelings of favoritism and unfairness, in which subordinate units located in different states from their parent organizations believe that they are discriminated against in favor of in-state subordinates, who as a result enjoy deployment opportunities, priority for new equipment, and priority for training courses that do not accrue to the out-of-state unit.

Title 32 Chain of Command

When units are not mobilized for federal service, they operate under the authority of Title 32, USC. Figure 2.5 illustrates the basic relationships. As the figure indicates, the lines of authority run from the Secretary of the Army down to the Director, Army National Guard, within the National Guard Bureau. This line of authority extends to the state adjutant general, who functions as the senior military officer within the state. As the left side of the chart indicates,

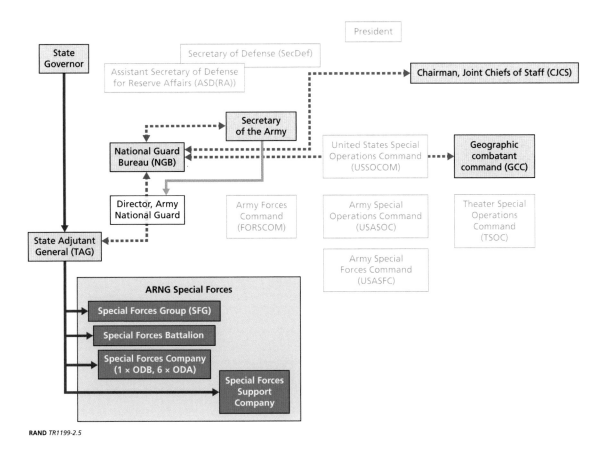

Figure 2.5
Title 32 Chain of Command

RAND *TR1199-2.5*

there is a second line of authority that descends from the state governor to the adjutant general. This line in the chain of command supports the governor's use of the National Guard in times of disasters, emergencies, and insurrection. Note that, at the bottom of the chain, the ARNG Special Forces are simply depicted in a green box.

This illustration is meant to represent the variety of different relationships with other forces the ARNG Special Forces can have, depending upon their state. Some states with a Special Forces headquarters, typically a group or battalion, will subordinate companies and detachments to that headquarters. In other instances where a state has a single Special Forces company (e.g., Rhode Island), it is likely to be subordinated, along with other small units, to some larger unit for administration and military justice. Other states, such as Texas, which has two companies from different battalions of the 19th SFG, subordinate both companies to an airborne infantry battalion while in Title 32 status.[5] Under Title 32, there is very little coherence in command relationships for Special Forces units.

Title 10 Chain of Command

When ARNG units are mobilized for federal service, they are subject to the federal chain of command under the authority of Title 10, USC. Figure 2.6 illustrates the basic relationships.

---

[5] Interview with LTC Douglas K. O'Connell, Texas ARNG, January 17, 2011.

Figure 2.6
Title 10 USC Chain of Command

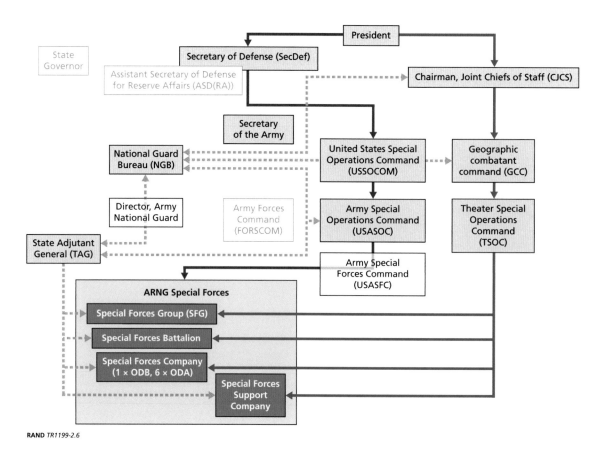

RAND *TR1199-2.6*

Once mobilized, the ARNG Special Forces are subordinated to USASFC as illustrated in the figure. Like AC Special Forces, they can then be deployed anywhere, usually assigned under the authority of a theater special operations command (TSOC) as suggested by the line of blue arrows down the right side of the figure.

1095 Rule

According to a memorandum issued by the National Guard Bureau,[6] National Guard members are restricted to a maximum of 1,095 days active duty special work (three years) out of the preceding 1,460 days (four years). Discussions of the rule with former OSD officials indicate that it is more of an accounting tool rather than an operational constraint. A waiver is available from the Director, Army National Guard, in order to exceed 1,095 days.

---

[6]   Departments of the Army and the Air Force National Guard Bureau, "Subject: Active Duty Special Work (ADSW) Title 10 Guidance," September 30, 2005.

ARNG Special Forces Supply

ARNG Special Forces units, almost without exception, are manned above 100 percent. The constraint on the supply of ARNG Special Forces is not manpower, but the percentage of those personnel who are qualified in the military occupational specialty associated with their duty position in the organization (duty MOS qualification, or DMOSQ in Army jargon). Figure 2.7 summarizes trends in DMOSQ.

The y-axis reflects the percentage of personnel qualified in their duty MOS, in this case, the 18-series that identifies Special Forces personnel. The blue line reflects the qualification rate among warrant officers, which may seem counterintuitive given the central role that warrant officers play in Special Forces. We understand that in the ARNG, a number of warrant officers joined Special Forces units from other career fields (e.g., ordnance, aviation) and have not yet attended the Special Forces Qualification Course. The trend in commissioned officers, reflected in the red line, results from the same factors, where officers qualified in one MOS move to Special Forces and must become qualified as 18-As. Of particular importance is the DMOSQ percentage for O-3s. A FY 2010 "Health of SOF" report indicates that the lowest DMOSQ percentages for commissioned officers are for O-3s, and in fact, O-3s are only behind CW3s for the lowest DMOSQ percentage in the ARNG SFGs. This is critical due to the fact that O-3s are in command of the ODA, the most basic building block for SF. Enlisted MOS qualification rates, illustrated by the black line, are somewhat higher still.

As the figure indicates, the trend has been positive, but none of the ARNG Special Forces Groups has yet attained the desired 85 percent DMOS qualification rate.[7] Given these circumstances, ARNG Special Forces units must "cross-level" (add MOS-qualified soldiers from other units in order to man a fully qualified unit for deployment). In the judgment of Army

Figure 2.7
Recent Trends in Duty ARNG Special Forces MOS Qualification Rate

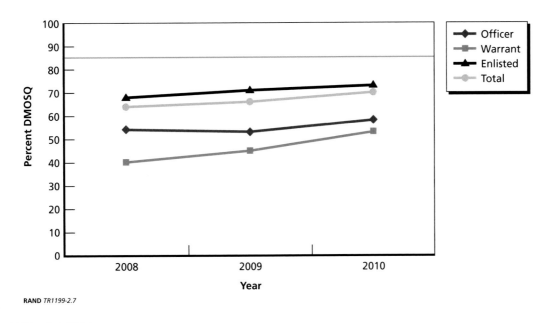

RAND TR1199-2.7

---

7   In accordance with USASOC Regulation 350-1.

National Guard officials and that of 19th and 20th Special Forces Group senior leadership, this has not posed a problem in generating appropriately trained units when called upon to do so. However, an AC colonel with direct experience in the chain of command for deployed ARNG Special Forces noted personality clashes, a lack of unity of effort within the cross-leveled unit, and self-serving behavior on the part of its members (e.g., going to military free fall school during post-mobilization training when there was no requirement for such skill on the deployment): behaviors he attributed to cross-leveling.[8]

Comparative General Characteristics of the Force

ARNG Special Forces soldiers attend the same MOS qualification training their AC counterparts attend. They likewise attend the same additional-skill identifier (ASI)-producing courses as their AC counterparts: most typically the military free fall parachute course, the dive ("scuba") course, and sniper course. They also attend the advanced special operations training (ASOT) and attain high-level (CL III) qualifications there. We use the statistics for these qualifications along with age and average years of service as the basis for a comparison of the general characteristics of the ARNG and AC Special Forces in Table 2.1, below. The table reports National Guard data on the basis of information from the two SFGs. The AC data reflect information based upon the five active groups. The years of service and age columns confirm that ARNG soldiers are slightly older and have served somewhat longer, although not all of that service was on active duty in the case of the ARNG. The table reflects "normalized" figures for military free fall, dive, and sniper qualifications.[9] We took the total number of qualified soldiers reported in each component (AC and ARNG) and divided it by the numbers of SFGs in their force structures to reveal the nominal density of skills per group. As the table illustrates, the AC groups have almost a four-to-one advantage in military free fall, nearly a three-to-one advantage in dive, and the ARNG has a 1.45 advantage over the AC in sniper training. ASOT qualifications are not included because they are classified.

Language Qualification

Traditionally, language qualification has been an important attribute of Special Forces. The demands of multiple deployments into primarily two countries have taken their toll on lan-

Table 2.1
Comparison of General Characteristics of the Force

U.S. Army National Guard	Years of Service	Average Age	Military Free Fall	Dive	Sniper
Enlisted	15	36			
Warrant	22	43	70/SFG	42/SFG	121/SFG
Officer	16	37			
Active Component					
Enlisted	12	33			
Warrant	19	36	267/SFG	119/SFG	83/SFG
Officer	13	39			

NOTE: Individual skills reflect best-available but incomplete data.

---

[8] Telephone interview, January 5, 2011.

[9] Although there are many more ASIs, these are the most abundant in the force.

Table 2.2
ARNG Language Qualifications

Reading Proficiency							
Level	0	0+	1	1+	2	2+	3
19th SFG	4	6	19	6	34	26	26
20th SFG	85	105	71	38	44	29	38
Languages Reporting Numbers of Readers							
		Numbers of readers	1 to 5	6 to 10	11 to 15	16 and more	
		19th SFG languages reporting	15	1	3	2	
		20th SFG languages reporting	12	3	0	3	

guage skills in both the AC and ARNG groups, we are told. ARNG language skills have also been undermined by realignments on different geographic regions and changes in directed training alignments with the AC SFGs, which in some instances have required ARNG soldiers to abandon one language and pursue a new one. This section of the chapter reports on ARNG language proficiency. Table 2.2 summarizes the data.

The table uses reading proficiency as a measure of language qualification. The Defense Language Institute measures language proficiency along a scale from "0" through "3," with "+" indicating additional proficiency not warranting award of the next-higher number. The table reports the numbers of soldiers in each SFG reading at each proficiency level. Thus, the reading proficiency part of the table shows that 24 percent of 19th SFG scored in the lowest proficiency categories (0, 0+, and 1) while 43 percent scored in the highest two proficiency categories (2+ and 3). In 20th SFG, 64 percent scored in the lowest three categories (0, 0+, and 1) while 16 percent scored in the top two (2+ and 3).

The lower half of the table presents a different perspective on language proficiency by reporting the number of readers associated with any given language. Thus, a language reporting readership between 1 and 5 has very low density within the unit, while languages at the other extreme, those reporting 16 or more readers, demonstrate some depth. As the table shows, neither group has much depth in any language. The 19th SFG has 16 or more readers in only two languages, and 20th SFG has 16 or more readers in only three.

Some of the ARNG Special Forces members we spoke with indicated there was a certain cynicism about language training that remained as a residual effect of past reorientations and realignments toward other geographic regions with other language needs. They also indicated that it was easier to motivate dedicated language training when a deployment loomed and the soldiers could see the need for real language proficiency. Figure 2.8 provides an imperfect snapshot of current language qualifications. Spanish is excluded from the figure for readability reasons. There are nearly 350 ARNG Special Forces soldiers who have Spanish language scores.

Deployment History

The deployment history of ARNG Special Forces is especially salient to a discussion of the supply of these forces. The frequency of deployment speaks to several issues: first, their current availability—is there any remaining capacity, or has the ARNG reached some limit that

Figure 2.8
Snapshot of Language Qualifications (Spanish Excluded)

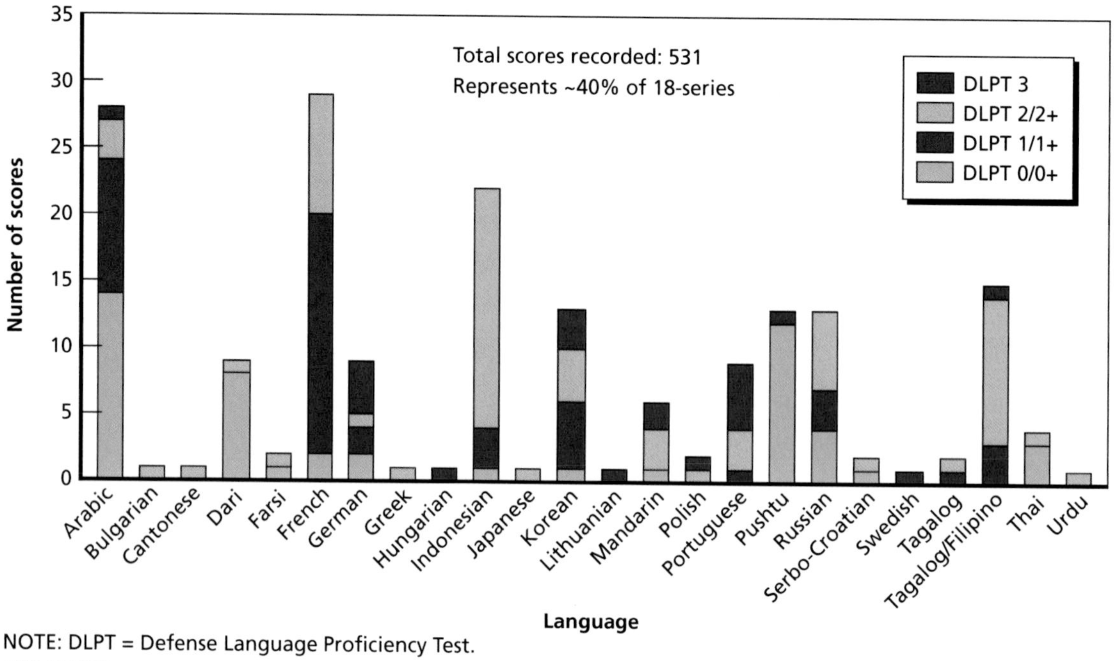

NOTE: DLPT = Defense Language Proficiency Test.
RAND *TR1199-2.8*

precludes additional deployments (e.g., a BOG:dwell redline[10])—and second, their experience and competence relative to the AC force. After nearly 10 years of war this latter consideration is a potentially significant one. Given that the AC force is very experienced and considers itself highly competent, interviews with AC members indicate that it worries about what additional risk it incurs if it shares operations with a less seasoned, less competent partner—although they accept this risk routinely when training and operating with indigenous forces. Figure 2.9 reflects the current deployment status.

The average number of deployments for MOS 18-series personnel reflected in the figure is 1.36. Over 500 soldiers have deployed once, and nearly 400 have deployed twice. Approximately 470 have never deployed, although this is typical of ARNG units in general, because unlike the AC, they have no trainee, transient, holdee and student (TTHS) personnel account; they must carry these soldiers on their unit rolls. Another contributor to a large number of yet-to-deploy soldiers are new arrivals, soldiers who have joined a unit during the time many of its members are deployed and who will have to wait until the unit mobilizes for the next deployment before they get their chance.

Figure 2.10 reflects the average number of deployments for the non–Special Forces personnel (i.e., the "enabler" MOSs (military intelligence, quartermaster, medical, signal, etc.)) in the 19th and 20th SFGs.

The average number of deployments in the figure is 0.74. There is an abundance of support personnel who have never deployed. The same explanation for the nondeployers offered

_____

[10] The ratio of boots on the ground or BOG (time deployed) versus "dwell" (time at home) has become the standard metric for the frequency and intensity of employment of both AC and RC forces. A "BOG:dwell redline" means a frequency and intensity of deployment that military personnel find unreasonable.

Figure 2.9
Current Deployments of ARNG Special Forces Personnel

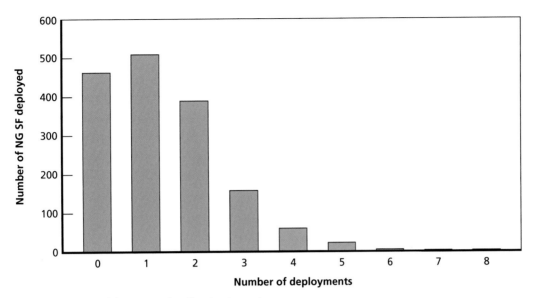

NOTE: Data derived from records reflecting hazard pay..
RAND *TR1199-2.9*

Figure 2.10
Deployment of Non-18-Series MOSs from 19th and 20th SFGs

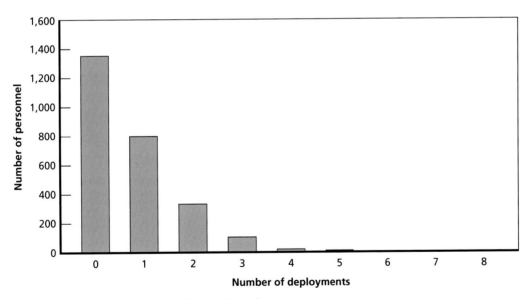

NOTE: Data derived from records reflecting hazard pay.
RAND *TR1199-2.10*

above explains the numbers of nondeployers in this category. After calculating the average MOB:dwell ratio (the ratio reflecting the amount of time soldiers are mobilized in federal service relative to the amount of time they spent at home station in Title 32 status) for the ARNG Special Forces, we asked our interviewees whether they viewed it as a constraint (because it reflects a far smaller proportion of time at home to time mobilized/deployed than Secretary Gates' 1:5 guideline for the RC). We were told nearly unanimously that it was not a constraint. This sentiment was confirmed by data in the Special Forces survey, where 84 percent answered "yes" to the question "If circumstances were right would you volunteer to deploy as an individual to fill a slot in an AC SF or to perform some other specified Special Forces function?" Likewise, when asked "How frequently would you be prepared to deploy," 62 percent indicated "one out of three (years)"; an additional 15 percent replied "one out of four." We conclude from this that the ARNG Special Forces retain the willingness for additional deployments.

Individual deployments can be different from unit rotations. Figure 2.11 indicates the unit mobilizations and deployments for ARNG Special Forces units. The figure indicates that except for the second half of FY 2004 and the first half of FY 2005, ARNG SFGs generally have at least portions of two or three units deployed at any one time.

Experience and Competence

Now that the RC has become an operational part of the United States' nearly decade-long campaigns in multiple theaters, there are expectations of experience and competence. Units and personnel who have not yet deployed face profound skepticism in meeting such expectations. Figure 2.12 illustrates the dynamics of the problem.

The figure illustrates notional competence on the vertical axis, and notional time in years on the horizontal axis. The red dot closest to the y-axis marks the point in time when two officers, one ARNG, the other Regular Army, graduate the Special Forces Qualification Course together. At that point their experience and competence levels are nearly identical. Subsequently, the Regular Army officer deploys multiple times, as indicated by the green dots moving up and to the right. With each deployment, he becomes more competent as a Special Forces member. The career of the ARNG officer reflects the same dynamics, but because his

Figure 2.11
ARNG SFG Mobilizations, FY 2002–2010

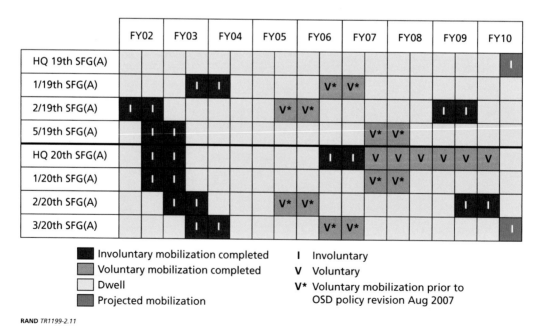

	FY02		FY03		FY04		FY05		FY06		FY07		FY08		FY09		FY10	
HQ 19th SFG(A)																		I
1/19th SFG(A)					I	I			V*	V*								
2/19th SFG(A)	I	I							V*	V*						I	I	
5/19th SFG(A)			I	I							V*	V*						
HQ 20th SFG(A)		I	I						I	I	V	V	V	V	V	V		
1/20th SFG(A)		I	I								V*	V*						
2/20th SFG(A)			I	I			V*	V*								I	I	
3/20th SFG(A)				I	I				V*	V*								I

■ Involuntary mobilization completed   I   Involuntary
▨ Voluntary mobilization completed   V   Voluntary
□ Dwell   V* Voluntary mobilization prior to
▨ Projected mobilization   OSD policy revision Aug 2007

Figure 2.12
Deployments as a Factor in Building Competence

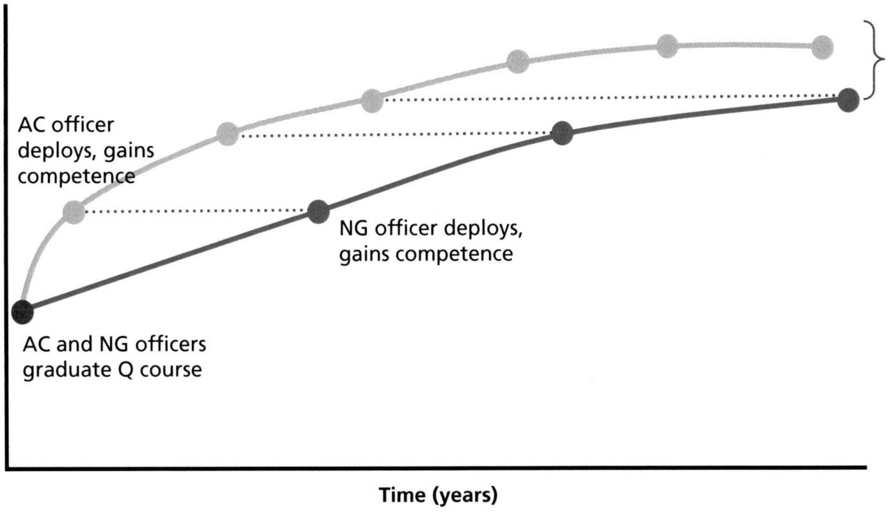

RAND *TR1199-2.12*

deployments occur (by design) less frequently, his growth in competence takes a different trajectory, and may even experience some decay if he is not given the opportunity to deploy at a frequency suitable to maintain his skills. Thus, deployment has become an essential part of the ARNG Special Forces pedigree; they must deploy in order to maintain and build their operational competence.

There may have been a time when RC units and personnel served as the strategic reserve and where the absence of deployed experience was not, by itself, a cause for concern. According to interviews with both AC and ARNG Special Forces personnel, those days are gone. With the ARNG deploying as an integral part of today's military operations, to deny units and personnel deployment opportunities is to render them untrustworthy in the eyes of the rest of the force—including their ARNG peers who have deployed.

So, how are ARNG Special Forces perceived currently? Figures 2.13 through 2.17 reflect responses from the Special Forces survey conducted as part of this project. Figure 2.13 captures the AC Special Forces respondents' assessment of ARNG Special Forces ability to operate ODAs. The y-axis reflects the percentage of respondents. The x-axis reflects the response choices the survey presented. Respondents could characterize the ARNG's ability to operate ODAs as limited, about the same (as AC ODAs), more (effectively than the AC), and no response. Each column is color-coded to reflect the respondents' frequency of involvement with the ARNG ODAs on the notion that the frequency of involvement with them produces sounder assessments. It is therefore important for readers to pay attention to respondents reporting daily and weekly involvement, because these responses may be more highly informed than the views of respondents shaped by less frequent involvement.

In the case of Figure 2.13, approximately 47 percent of respondents assessed the ARNG ODAs as limited in their capabilities; this includes 30 percent of respondents reporting daily or weekly involvement with the ARNG. Approximately 28 percent of respondents indicated ARNG ODAs were equivalent to their AC counterparts. Twenty percent of these respondents had daily or weekly contact with the ARNG units in question. A minority of approximately 4 percent indicated they thought the ARNG ODAs were more capable than their AC counterparts, but nearly half of the respondents did not report the frequency of their involvement with the units in question. Finally, 20 percent of those surveyed offered no response to the question.

Figure 2.14 tells a similar story about AC views of the capabilities of ARNG ODBs. About 37 percent of respondents indicated they thought the ARNG ODBs provided limited

Figure 2.13
The AC Special Forces View of ARNG ODAs

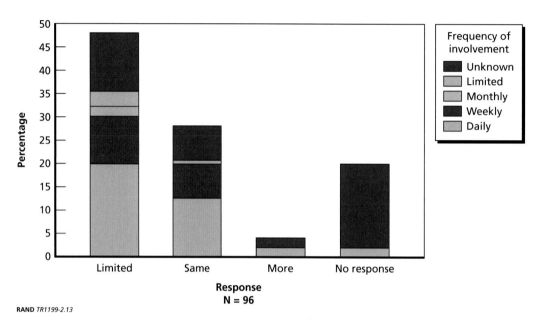

RAND TR1199-2.13

Figure 2.14
AC Special Forces Assessment of ARNG ODBs

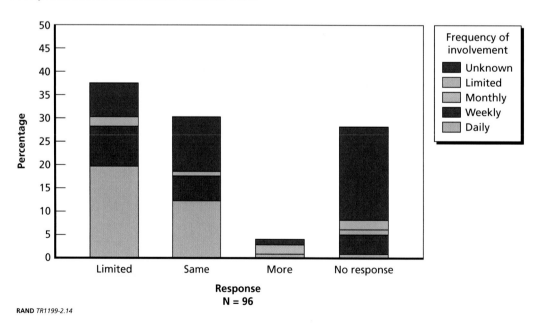

RAND TR1199-2.14

capabilities, and approximately 28 percent of respondents holding this view reported daily or weekly involvement with the ARNG units.

About 30 percent of respondents assessed the ARNG ODBs as the same as their AC counterparts, with about half the respondents reporting daily or weekly involvement with the units in question. Less than 5 percent of respondents believed ARNG ODBs were more capable than their AC counterparts. Some 27 percent of survey participants offered no answer to this question.

Figure 2.15 summarizes the AC view of the ARNG's ability to operate SOTFs and JSOTFs. The figure tells a by now familiar story. Approximately 38 percent of respondents rated the capabilities of these ARNG units as "limited." Approximately 27 percent of those holding that view had daily or weekly involvement with the units in question. Approximately 37 percent of respondents assessed the ARNG units as the same as their AC counterparts, with about 21 percent reporting daily or weekly contact with the units in question. None rated the ARNG as more effective than the AC. Approximately 26 percent of those participating in the survey failed to answer this question.

Finally, Figure 2.16 presents the AC Special Forces view of the ARNG Special Forces capabilities in terms of combat support (CS) and combat service support (CSS). Here, the assessments are somewhat more positive. Only about 23 percent of respondents concluded that the CS/CSS capability of ARNG units was limited compared to that of the AC, and approximately 14 percent of the respondents reported daily or weekly involvement with these units. Approximately 46 percent of respondents rated ARNG Special Forces CS/CSS as the same as that in the AC: a significant finding if one considers that the ARNG groups have a general support company to provide their CS and CSS, while the AC groups have battalions. Finally, one-third of survey participants failed to answer this question.

The ARNG Special Forces have their own view of their capabilities relative to that of their AC counterparts, presented in Figure 2.17. A significant percentage of respondents believe their

Figure 2.15
AC Special Forces Assessment of ARNG SOTF/JSOTF

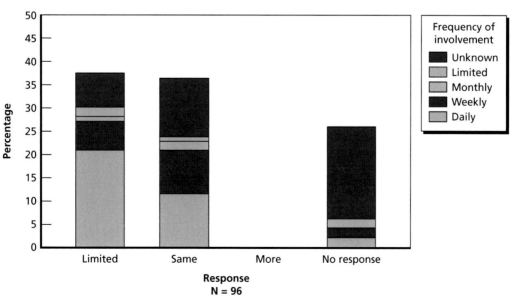

RAND TR1199-2.15

Figure 2.16
AC Assessment of ARNG Special Forces CS/CSS

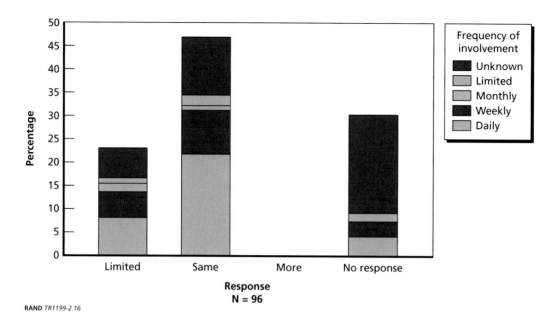

RAND *TR1199-2.16*

Figure 2.17
ARNG Assessment of Its Capabilities Relative to the AC

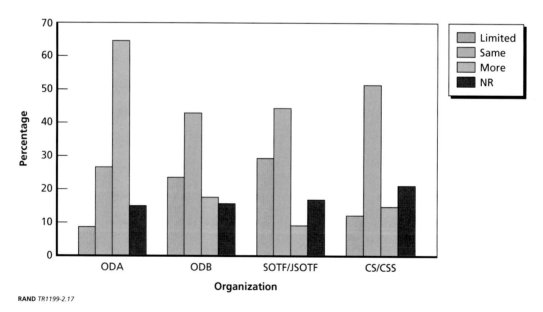

RAND *TR1199-2.17*

capability at ODA level is the same as or more effective than that found in AC ODAs. The approximately 64 percent reporting "more" capabilities is consistent with interviews in which the ARNG interviewees pointed to their age, maturity, civilian background, and depth of experience as factors that made them more effective at ODA level. The roughly 9 percent who assessed ARNG ODAs as limited compared to AC ODAs is also consistent with interviews in which members of the ARNG SFGs acknowledged that they do not "get the repetitions" (i.e., do not get to employ their ODA-level skills as frequently as the AC units) that the AC units do. These interviewees usually believed that they could close the gap during post-mobilization training, however.

The self-assessment of ODB capabilities is also in line with what we heard in interviews, which generally tended to rate ARNG ODBs as equivalents to the AC units. Figure 2.17 indicates that approximately 24 percent of respondents believed ARNG ODBs were limited in their capabilities relative to the AC units of the same type. About 18 percent, however, held the opposite view, indicating that ARNG ODBs were more capable than the AC units. About 43 percent stated that ARNG ODBs have the same capabilities as an AC ODB.

The way in which the "limited" assessments grow from ODA responses to SOTF/JSOTF is consistent with the message that emerged from many discussions with ARNG Special Forces. They generally were most confident in their abilities at ODA and ODB level, and acknowledged the challenges, especially in complex tasks and command and control associated with SOTFs and JSOTFs. The CS/CSS ratings seem consistent with the confidence most ARNG Special Forces members expressed in their "enablers."

Women in Supporting Organizations

One of the questions to emerge from USASOC concerning "enablers" had to do with women and their relative abundance in the ARNG SFG general support companies. The question seemed to explore whether there are enough women present to have utility in interfacing with women in indigenous populations where it might be inappropriate for men to do so. Table 2.3 summarizes the total number of women currently assigned to the 19th and 20th SFGs, along with their branches of service and MOSs.

Table 2.3
Women in the ARNG Special Forces Groups

Branch	MOS	Number
Unqualified	NA	9
Unmanned Aerial Systems Operator	15W	1
Multimedia Illustrator	25M	1
Judge Advocate	27A	1
Intelligence Analyst	35F	3
HUMINT Specialist	35M	4
Human Resource Specialist	42A	7
Financial Management Specialist	44A	1
Emergency Physician	62A	1
Health Services Administrator	70B	2
Environmental Science and Engineering	72D	1
Chemical, Biological, Radiological, Nuclear Operations Specialist	74D	1
Motor Transport Operator	88M	1
Wheeled Vehicle Mechanic	91B	1
Food Service Specialist	92G	1
Parachute Rigger	92R	3
Water Treatment Specialist	92W	1
Unit Supply Specialist	92Y	1
Computer Detection Systems Repairer	94F	1
Total		41

As the table suggests, the women have a variety of MOSs, and are few in number: 41 total in the two groups. Nine of the soldiers reported in the table are not yet MOS-qualified and are likely new members. The skills reported suggest that some of these women are assigned to the group headquarters company, and others to the general support company.

Having characterized the supply of ARNG Special Forces (and their availability for future deployments) and offered judgments about their capabilities relative to those of the AC Special Forces, the next subsection turns to consider the civilian skills, knowledge, and capabilities resident within the force.

ARNG Civilian Skills, Knowledge, and Capabilities

In almost every conversation or interview we conducted in the course of this study, someone asserted the value of civilian skills, knowledge, and capabilities residing within the ARNG Special Forces. In designing the Special Forces survey, therefore, we included questions that asked the respondents to identify their civilian skills.

There is a general expectation, held by ARNG members and implicit in LTG Mulholland's expectations of niches, that the civilian careers of Guardsmen provide them with a wider range of experiences than their AC counterparts typically enjoy, and that these experiences may make ARNG Special Forces soldiers more capable of relating to indigenous persons on an individual level. Taken a step further, it might be that the diversity of civilian background itself (as opposed to the specific careers and skills examined below) is a valuable attribute of the ARNG Special Forces soldier.

Civilian Skills in the Force

From a total of 525 respondents, 418 answered the question about their civilian skills. They self-identified 53 separate skills, careers, or capabilities. Figure 2.18 reports the percentage in each of the top career/skill fields. If we assume the universe of ARNG personnel in the 19th and 20th SFGs is 4,000, then the survey provides a 13 percent response rate. If we further assume that respondents and nonrespondents from all civilian career fields would be equally likely to answer the civilian skill question, we can use the survey responses as a basis from which to extrapolate rough estimates of the total number of each career or skill present in the force.[11] Table 2.4 provides the extrapolated estimates.

Some skills are very low density. Fourteen of the 22 skills reported are estimated to contain fewer than 100 men each. It becomes, therefore, very difficult to make systematic use of most of the skills displayed, with the possible exception of law enforcement and perhaps the medical category. The anecdotes shared with us about ODAs filled with builders or police are probably best understood as the results of uneven distribution of these skills that form concentrations in certain locales or within certain Special Forces units. Their contributions have certainly been important, but probably very difficult to promote throughout the force because of the uneven distribution of these skills. A final point is that individuals may volunteer to be

[11] There is limited research with military audiences on the topic of nonrespondents, but early research shows nonrespondents do not differ all that greatly from respondents. See Carol E. Newell, Paul Rosenfeld, Rorie N. Harris, and Regina L. Hindelang, "Reasons for Nonresponse on U.S. Navy Surveys: A Closer Look," *Military Psychology*, Vol. 16, No. 4, 2004, pp. 265–276.

Figure 2.18
Percentages of Career/Skill Density Within ARNG Special Forces Among Survey Respondents

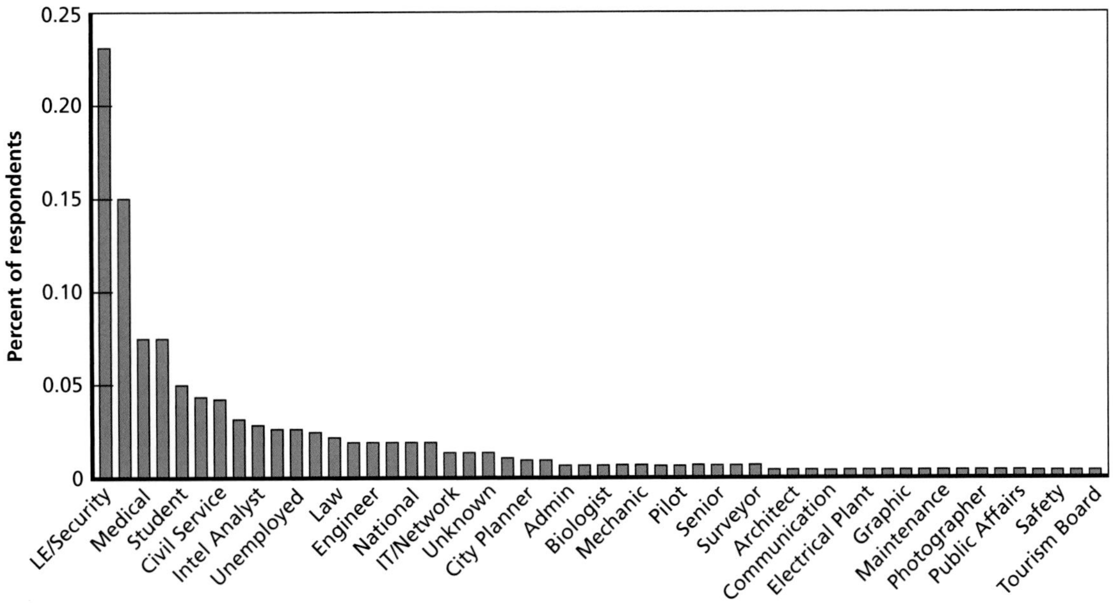

RAND *TR1199-2.18*

Table 2.4
Extrapolated Estimates for Numbers of Personnel of Given Skills in the Force

Career or Skill	Number of Respondents in Survey	Estimated Number of Soldiers with It
Law Enforcement/Security	91	1,040
Medical	29	320
Small Business	12	240
Utility Rate Analyst	17	240
Trades	29	200
Student	19	200
Defense Contractor	9	160
IT/Network	7	160
Firefighter	10	120
Civil Servant	16	120

mobilized based on requests for their civilian skills, but they cannot be involuntarily mobilized based upon their civilian skills.

One potentially important civilian attribute was not captured in the survey. In discussing what they perceived as key elements of their civilian backgrounds that enhanced their capabilities as Special Forces soldiers, many ARNG Special Forces interviewees pointed to the fact that most of them lead predominantly civilian lives in which they cannot rely on a chain of command or published priorities in order to be successful. They learn by experience the importance of negotiation, accommodation, compromise, persuasion, and other social skills. It is these skills many believe give them an edge over their AC counterparts in nonkinetic activities, foreign internal defense (FID), and some aspects of unconventional warfare (UW).

Demand for Special Forces

The demand for Special Forces is significant,[12] and represents both an opportunity to deploy ARNG Special Forces to grow their operational competence and the confidence that the AC force has in them, and the imperative to satisfy specific requests for forces. Without drawing upon classified information, this section of the chapter considers three aspects of current operations that affect the demand for Special Forces.

ARNG as Operational Reserve, Integrated into the Playbook[13]

As noted earlier in the supply discussion in this chapter, the average number of deployments for ARNG Special Forces personnel is 1.36; they are deploying and contributing to the current operations in Afghanistan, Iraq, the Horn of Africa, and elsewhere. Each of these theaters could experience a spike in activity that might prompt the regional commander to request additional forces. As concluded in the supply discussion, the ARNG Special Forces have additional capacity and could be employed to help satisfy future episodic spikes of violence (e.g., a terrorist surge). In addition, Operation New Dawn[14] is in its infancy, with uncertain potential to change the demand in terms of numbers of forces and types of missions. ARNG Special Forces could serve as a shock absorber to mitigate stress on the AC part of the force resulting from Operation New Dawn.

Theater Security Cooperation Demand

Theater security cooperation is among the fastest-growing sources of demand. The Guidance for the Employment of the Force (GEF)[15] elevated security cooperation and its related activities to new importance as a tool for advancing and defending U.S. interests. Since then, COCOMs have published their theater security cooperation plans and the services their security cooperation support plans. Together, these new planning efforts have fostered ambitious agendas with U.S. partners around the world, and many of them are suitable for Special Forces. Moreover, in FY 2011, only 55 percent of theater security cooperation activities have been resourced. Many of these typically require ODAs and sometimes ODBs. Still others, like Special Operations Command Central–Horn of Africa and Operation Enduring Freedom–Philippines, often require higher-level formations including SOTFs, JSOTFs, SFG headquarters, and similar formations, but rarely do they require a complete Special Forces battalion HHC or Group HHC.

Also looming is the prospect of longer theater security cooperation activities, such as joint combined exchange training events (JCETs) that run longer than the standard 30 days. These might be opportunities for overseas training deployments.

---

[12] See Lolita C. Baldor, "US Special Forces Show Strain, Says Admiral," *Boston Globe*, February 9, 2011, p. 8.

[13] The "Playbook" is the USASOC management document that programs requirements to deploy special operations forces of various types and then assigns specific units, both AC and RC, to fill those requirements.

[14] The name of current U.S. operations in Iraq, i.e., post-Operational Iraqi Freedom.

[15] Office of the Secretary of Defense, 2008.

Contingency Operations and Emergency Responses

Theater security cooperation events and activities largely represent scheduled demand: events known and mutually agreed to with specific partners. Contingency operations and emergency responses constitute the unscheduled portion of demand. This part of demand is more difficult, though not impossible, for the ARNG Special Forces to handle. That said, from a demand management perspective, it is probably prudent to maximize the ARNG's role in scheduled demand in order to provide maximum predictability and to exploit the inherent responsiveness and agility of the AC to respond to contingency and other unscheduled demands.

The USASOC Playbook and the ARNG Unit Life Cycle Interactions

Figure 2.19 begins the discussion of the relationship between USASOC and the ARNG Special Forces by roughly describing the Playbook sequence of events and the ARNG unit life cycle of alert, mobilization, deployment, release from federal active duty, and return to state duty. The Playbook sequence unfolds when U.S. Special Operations Command (USSOCOM) announces the future force requirements to its service component commands (e.g., USASOC and U.S. Air Force Special Operations Command). USASOC is responsible for resourcing the Special Forces, Ranger, and Army Special Operations Aviation requirements. To do so, it holds force generation conferences to assign units against requirements. USASFC then creates the Playbook, indicating what SF units will deploy, when, and to what theater for what purpose. Once the contents of the Playbook have been approved, the Joint Staff publishes the deployment orders and the units deploy as directed. One should realize the Playbook focuses primarily on SFG HHC and battalion responsibility for overseas contingency operations.

Meanwhile, in the ARNG, units released from federal active duty (REFRAD) return to their armories and their state chains of command, where all ARNG units are under the authority of the state adjutant general. They spend their training time maintaining basic, often individual, military skills. For ARNG Special Forces, this typically involves working on language

Figure 2.19
The USASOC Playbook and the ARNG Life Cycle

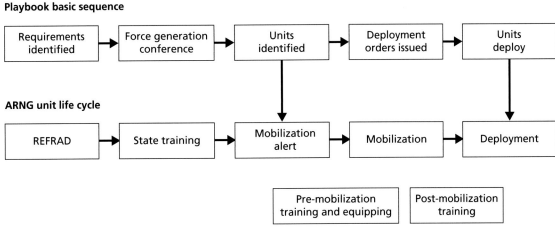

skills in the unit language lab, maintaining their parachute qualification with a jump at a local airfield, or undertaking small arms or sniper training at a local range complex.

ARNG units are supposed to receive 24 months of alert prior to mobilization, and no less than 90 days.[16] Force generation conferences therefore must be planned at such a frequency so as to provide this alert. In practice, the group commanders from the two ARNG SFGs have not always been invited to the force generation conferences. Decisions about mobilization of their subordinates have been made without their input. Moreover, the force generation process, especially without participation by the ARNG SFGs, lacks a rigorous process for making dispassionate decisions about the next to deploy. This opaque process fuels a sense of grievance among the Guardsmen, along with suspicions that the ultimate decisions are heavily influenced by mistaken AC perceptions of the Guard.

At some point after an ARNG unit is notified that it will mobilize for federal service, it begins its pre-mobilization training and begins receiving the equipment it will need during its deployment. This is a critical point in each unit's preparations, and a time when the unit is largely alone for the reasons given earlier in this chapter. It has not yet mobilized, so the remainder of its active duty (Title 10) chain of command is not yet operable; yet the ARNG Special Forces unit is often the sole Special Forces entity within its state. Its adjutant general and other state units can offer it little in terms of support for its pending mobilization.

Once the unit mobilizes and becomes a federal asset, it undergoes post-mobilization training to prepare it for the conditions it is likely to face once deployed. Much of this training is theater specific and required for deployment, but does take away from collective training. For example, units get mine-resistant, ambush-protected (MRAP) training if they are going to a theater where improvised explosive devices (IEDs) are a significant threat. Many individuals interviewed also noted that a significant amount of time was spent on new equipment training, such as communication gear. These training requirements compete with the unit's collective training requirements. At some point in this process, the unit links up with the AC unit it will accompany. Alternatively, the unit may deploy alone.

U.S. Army Special Forces Command developed a new mobilization policy in 2010 that is expected to overcome many of the obstacles and concerns described here. In addition, there is broader effort to create a SOF force generation process akin to the Army Force Generation cycle used by general purpose forces for both the active and reserve components. The point remains, though, that there are critical dependencies between the Playbook and the USASOC force generation process on one hand and the ARNG unit life cycle on the other. In order to assure getting appropriate resources from the ARNG, USASOC's force generation procedure should be rigorous, transparent, and routinely involve the ARNG SFG senior leadership.

Chapter Conclusions

The following points emerge as central from our discussion of the problem space for ARNG Special Forces:

---

[16] Some 19th SFG personnel have deployed with as little as 30 days warning.

- There are a few, but not many, statutory and policy issues that interfere with USASOC's ability to maximize the contributions of ARNG Special Forces or move them toward a "purpose-driven force."
- In particular, DoDI 1235.10 (Enclosure 2) casts the ARNG as an asset of last resort, and although this characterization is apparently not treated as authoritative, it might usefully be rewritten to reflect the type and intensity of integration DoD expects from the AC and the ARNG.
- The current chain of command provides limited functionality before mobilization and has prompted widespread, though not universal, suspicions of unfairness and favoritism.
- Supply of ARNG Special Forces is ample in terms of manpower, but below goal in terms of duty MOS qualification.
- Despite the limitations in terms of duty MOS qualification, the ARNG Special Forces contains additional capacity and will to undertake future deployments.
- The ARNG Special Forces generally have fewer personnel qualified in additional skills (e.g., military free fall, dive), than the AC, although they have more qualified snipers than the AC and the ARNG members typically are somewhat older and have more years of service than their AC counterparts.
- Language skills, a challenge throughout the force (both active and reserve), do not appear to be a principal asset of the ARNG Special Forces.
- Deployments are not only important to accomplish the mission, but are also essential for growing and maintaining competence within ARNG Special Forces and for winning the confidence of the AC.
- The prevailing view in the AC of the ARNG Special Forces is guarded, but far from dismissive. Conditional consensus appears to exist for the premise that there are tasks and circumstances suitable for the ARNG, especially at ODA, ODB, and CS/CSS unit levels.
- Both women in the ARNG Special Forces Groups and civilian skills of the members, while important, do not appear to hold the additional potential benefits that USASOC may have hoped for.
- Demand for Special Forces is high, though not at its apogee, and unstable. It appears that ARNG Special Forces can play a useful role in satisfying demand, especially demand that takes the form of scheduled events: programmed rotations within the Playbook and theater security cooperation activities.

Just what specific events and tasks present themselves as ideal for ARNG Special Forces? And what of the niches that USASOC seeks? These are the substance of the next chapter.

Strong Suits and Niches: Identifying and Playing to the Strengths of ARNG Special Forces

This chapter draws on policy-level interviews[1] and discussions during site visits with Special Forces personnel from both the AC and ARNG that the project team encountered in the course of the research. The policy-level interviews engaged subject matter experts at the assistant secretary of defense level, National Guard Bureau, state adjutants general, and uniformed officers down to the rank of colonel. Military officers interviewed came from both the ARNG and the AC. Interviews during site visits involved principally ARNG majors and captains, and senior NCOs in the ranks of sergeant first class through sergeant major. This chapter sifts through the collective insights from these various encounters—criticisms, observations, and recommendations—and tries to overlay them on the conclusions that emerged from Chapter Two in order to (1) endorse those conclusions and (2) amplify certain aspects of them in a way that supports the identification of options for making ARNG Special Forces purpose-driven. Their observations fall into seven broad categories, which provide the structure for this chapter:

- a need for greater mutual understanding, cooperation, and coordination
- the value of ARNG Special Forces
- constraints on the performance of ARNG Special Forces
- equipping issues
- options to enhance the contributions from ARNG Special Forces
- observations on the structure of the force
- niches for ARNG Special Forces

The Need for Greater Mutual Understanding, Cooperation, and Coordination

This general theme appears in several forms. First, there is a widely held view among the site visit interview participants that USASOC does not make the effort to understand its ARNG Special Forces: a circumstance several respondents found ironic, given the command's devotion to working "by, with, and through others."[2]

---

[1] Some officials answered in face-to-face interviews. Others preferred to fill out the study questionnaire. Still others preferred to discuss their answers before submitting them. And still others preferred to discuss their views on ARNG Special Forces contemporaneously.

[2] Site visit to 20th Special Forces Group headquarters, December 2, 2010, and to 19th Special Forces Group headquarters, December 20, 2010.

These site visit interviews also exposed the belief that there should be more ARNG Special Forces personnel at USASOC, and that a greater presence there would have mutual benefits. It would mean clearer representation of ARNG Special Forces issues within the command, it would provide the means to interpret ARNG issues for the command, and it would provide qualified National Guardsmen to support mutual policy development.[3]

A related sentiment held that ARNG advisors must be more aggressive in identifying issues and helping USASOC with their resolution. Advisors are too passive, according to this view, and do not provide the ARNG perspective or participate constructively with the command in resolving issues.[4]

Among policy-level interviews, there were several calls for better coordination between USASOC, the ARNG, and ARNG Special Forces. Several respondents asserted that USASOC and the National Guard Bureau should coordinate Title 10 functions more closely, especially program objective memorandum (POM) and pay and allowances (P&A) requirements. Others cited the need for a USASOC-sponsored mobilization conference that would refine the details supporting the Playbook and ensure that force employment decisions were informed by senior ARNG Special Forces leadership recommendations. Others prescribed restarting the annual SOF/Adjutants General conferences and affording the ARNG richer opportunities to shape the agenda.[5]

Two key issues complete this section. The first is the imperative for consistency and predictability; the second is the perception of favoritism that unfairly penalizes some ARNG Special Forces.

Consistency and Predictability

Site visit discussions with ARNG Special Forces almost invariably raised this point. They typically noted that the lengthy coordination they must go through to prepare their families for an upcoming deployment, notify their civilian employers of their pending absence, and the unit preparations for mobilization itself are elaborate and must coordinate many details in order for the unit to be ready to go on time. Consistency here is key, because their processes cannot handle last-minute delays or significant changes in operational details. The message was clear: if the AC wants the best the ARNG Special Forces can deliver, then it must set the tasks, conditions, and standards, and hold them constant whenever it is possible to do so.[6]

Predictability discussions struck a similar theme, specifically, that given notice that mobilization would occur at some future point, the unit could be ready. But the units all need lead time in order to prepare. ARNG units generally, and Special Forces are no exception, have a finite ability to manage uncertainty and to adapt to late-breaking new requirements.[7]

---

[3]   Site visits to 20th and 19th Special Forces Group headquarters and to Rhode Island and Texas 19th SFG subordinate units, January 11, 2011, and January 17–18, 2011, respectively.

[4]   Site visits to 20th and 19th Special Forces Group headquarters and to Rhode Island and Texas 19th SFG subordinate units, January 11, 2011, and January 17–18, 2011, respectively.

[5]   Policy-level interview responses from the adjutants general offices of Alabama, Utah, West Virginia, Florida, and Rhode Island.

[6]   Interviews at 19th and 20th Special Forces Group headquarters.

[7]   Interviews at 19th and 20th Special Forces Group headquarters.

If there were any one thing that USASOC could do to improve coordination and cooperation with its ARNG Special Forces, it would almost certainly involve improving consistency and predictability.

Favoritism

If high morale and confidence in one's leadership are force multipliers, suspicions of favoritism are force dividers. And, just as force multipliers enhance their units, dividers undermine theirs. Therefore, the suspicions of favoritism that loom among ARNG Special Forces personnel should be treated seriously. There are widespread suspicions that units situated within the home state of the SFG headquarters enjoy unfair priorities for mobilization, equipping, and schools. These sentiments seem to be most profound among outlying units, but there are similar concerns, more widely held, that pertain to relations with the AC.[8]

Suspicions concerning the motives of the AC take many forms. Some state-level officials characterized the Marine Corps SOF's deployment as having "stolen" their deployment.[9] But if there were an AC unit ready to go, why wouldn't it be the first choice? Some soldiers believed the rules were stacked against them, even to the point of depriving them of awards and decorations. One asserted he did not get "his bronze star" because they "hadn't gone kinetic" when, in his judgment, avoiding a firefight was a success story.[10]

However trivial these anecdotes may appear, there seems to be a genuine sense of grievance among the ARNG Special Forces that might undermine mutual understanding, cooperation, and coordination with the AC.

The Value of ARNG Special Forces

The more junior participants in site visit discussions tended to talk about the ARNG Special Forces' ability at the tactical level. They spoke about what they viewed as their strong suit (addressed below in the niches discussion), and their value in giving the AC units "a breather" or "a chance to take a knee" (i.e., to be a moderating factor for the AC's operations tempo).[11]

Many respondents, from adjutants general to team sergeants, made the case for the civilian skills and backgrounds of the ARNG Special Forces soldiers, and the value that brought to their employments abroad.

Some of the most senior officials we spoke with characterized the value of the ARNG Special Forces as consistent with the value of the ARNG generally: a backup force that handles essential but secondary tasks so the AC can concentrate on the most difficult tasks (a paraphrase of the conversation).[12]

---

[8] Site visit discussions with 19th and 20th Special Forces Group personnel, December 2010, January and February 2011.

[9] Response from the office of the Adjutant General, Utah.

[10] Site visit discussion with Rhode Island National Guard Special Forces personnel, January 11, 2011.

[11] Site visit discussions with Texas National Guard Special Forces personnel, January 18, 2011.

[12] Policy-level interviews with members of the Office of the Assistant Secretary of Defense for Reserve Affairs, January 25, 2011.

Constraints on ARNG Special Forces Performance

Our site visit discussions offered eight considerations that constrain, impair, limit, or interfere with the ARNG Special Forces ability to perform at their best. The most-cited by company-level leadership was that ARNG Special Forces require operational deployments to maintain and build their skills and to gain credibility with the AC, but that they cannot count on being deployed. They do not believe that, nearly 10 years into current operations, they can be credible without regular deployments under their belts. It follows that if they are not seen as credible, they won't be deployed, creating a vicious circle that could culminate in their irrelevance.[13]

The second major constraint is that there currently are not enough qualified ARNG Special Forces personnel to man cohesive teams.[14] Perspective matters with this consideration. Adjutants general look at the roughly 16 percent cross-leveling necessary to prepare Special Forces for deployment as a success story, given that they typically must cross-level at roughly 25 percent to man general purpose forces units.[15] The AC respondents may not have known the relative percentages of cross-leveling that typically occur, but nevertheless believe whatever it is, it is too great.

Mobilization lies at the heart of the next two constraints on ARNG Special Forces performance. Many with whom we spoke during site visits complained that mobilization was not standardized (and ought to be) and that USASOC does not support it.[16] They believe USASOC considers mobilization as a "Guard issue" and prefers to leave it at that ("let us know when you're sorted out"). During the course of this project, USASFC was developing a mobilization policy to deal with this issue. The second mobilization-related constraint is post-mobilization training, which many believe takes too long, and which keeps growing additional requirements (e.g., MRAP training). Even the less cynically minded respondents wonder about the additional requirements, which they have trouble connecting to their jobs once deployed, and suspect that these new requirements might be some subterfuge to prevent them from deploying.[17]

Then there is the time frame in which our respondents believe they can be ready to deploy. Most were confident that they could be ready with 90 days notice. Some believed they could be ready in 30 days under emergency circumstances.[18]

Getting ready to deploy is difficult at best, in the majority view. Typical problems include adequate billeting for training, administrative delays with arranging essential contracting (e.g., body armor technician support), unit credit cards, and administrative support. The units have

---

[13] Site visit discussions with Texas National Guard Special Forces personnel, January 18, 2011.

[14] As data presented in Chapter Two pointed out, neither the 19th nor the 20th SFG has MOS qualification rates at or above the required 85 percent level.

[15] Interview with MG Blalock, Alabama Adjutant General, December 2, 2010. For further discussion on reserve component unit stability, see Thomas F. Lippiatt and J. Michael Polich, *Reserve Component Unit Stability: Effects on Deployability and Training,* Santa Monica, CA: RAND Corporation, MG-954-OSD, 2010.

[16] Those interviewed mobilized at different times. The mobilization process and some issues have changed over time.

[17] U.S. Special Forces Command has recently published a new mobilization policy, which is expected to address many of these concerns.

[18] Site visits to 19th and 20th Special Forces Group headquarters.

no alternative but to struggle with these issues rather than to concentrate on arguably more critical operational preparations for the eventual deployment.[19]

Finally, there are two, soldier-level, quality-of-life concerns that potentially impact performance, according to our site visit discussions.[20] The first is the lack of family support groups, re-enlistment bonuses, and moving expenses. The second concerns the SOCOM Care Coalition. As noted, ARNG Special Forces units are widely dispersed across 18 states, and some 40 percent of members live out of state. When they deploy, their family members are often not within easy reach of military installations or even a unit rear detachment. Add to that their relatively limited familiarity with military pay and access to benefits, and these circumstances can be nearly overwhelming. To the deployed soldier, worry about the family's welfare can become a morale factor—just as it sometimes does among AC members, whose families typically enjoy better support. The absence of re-enlistment bonuses and moving expenses work to undermine the soldier's sense of being a valued asset. The SOCOM Care Coalition issue is that wounded National Guardsmen lose their benefits through the REFRAD process when they are well enough to leave the warrior transition unit where they have recuperated. Getting the benefits restored upon return to the home armory would be difficult under best of circumstances, but having to face the process after a lengthy recuperation, often far from home, is felt to be especially taxing.

Structure of the Force

Although many ARNG officials interviewed emphasized the importance of Section 104, Title 32, and asserted that it meant that ARNG Special Forces units should share exactly the same structure as the AC units,[21] this view was not borne out at assistant secretary of defense level. There, our interviewees held the view that unit structure—both in the AC and within the ARNG—is requirements-driven.[22] These two views are not necessarily inconsistent, especially under circumstances like today's, where ARNG and AC Special Forces units, including their enablers, relieve each other in the course of sustained rotations into and out of Afghanistan and other theaters. Rotations are easier to manage if the units involved are identical, and can replace each other on a one-for-one basis.

Equipping Issues

In policy-level interviews at assistant secretary of defense level, officials do not believe there are any issues with equipping ARNG Special Forces. The Army National Guard Comptroller

---

[19] Site visits to 19th and 20th Special Forces Group headquarters.

[20] Site visits to 19th and 20th Special Forces Group headquarters.

[21] Policy-level interview responses from the adjutants general of Alabama, Florida, Utah, and West Virginia.

[22] Policy-level interviews with members of the office of the Assistant Secretary of Defense for Reserve Affairs January 25, 2011.

indicates that there are no process problems.[23] Others noted that DoDI 1225.6, "Equipping," is being rewritten, and characterize that fact as favorable.

Members of the 19th and 20th SFGs raised two issues. First, some noted that state adjutants general do not understand MFP-11 (major force program) funding, which can cause difficulties if those funds become commingled with MFP-2 funds or are spent for general-purpose forces. Second, they note that equipment is often shipped to the parent unit rather than to the end-user unit. This practice may have made sense in the era of paper property books so that the new items could be brought under unit accountability. In the era of electronic media and property books, however, the practice makes no sense. Moreover, it causes problems because the parent unit does not have funds to forward the equipment to the end-user. Typically, the end-user unit must send personnel during a drill weekend to pick up the new equipment and bring it home.[24]

Some policy-level interview respondents offered equipping prescriptions. One was "buy seven rather than five of everything," meaning that USASOC should buy for the ARNG what it buys for the AC.[25] A more conservative prescription argued for buying current communications for the ARNG groups to ensure interoperability with the AC, while deferring other equipment upgrades until mobilization to manage costs.

Enhancing Contributions from the ARNG Special Forces

This section presents policy-level interview and site visit discussion respondents' views on a wide variety of steps that, if taken, could generally improve the contributions from the ARNG Special Forces. The final section collects the respondents' ideas about the operations for which ARNG Special Forces might be perhaps especially suited: the "niches" the project was tasked to identify.

Two propositions would require radical changes in the status of the ARNG Special Forces. One recommended moving the Special Forces from the ARNG and re-establishing them within the U.S. Army Reserve (USAR).[26] The advocates of this move believed it would produce easier access to Special Forces and better command and control (i.e., it would function across state lines at all times). The second proposal was to move the ARNG Special Forces under the direct authority of the National Guard Bureau rather than have them subordinated to state adjutants general. The first option is at least legally possible, although implementing it would run contrary to the 1994 distribution of Special Forces to the ARNG and Psychological Operations and Civil Affairs units to the USAR. The body of legislation and policy reviewed as a part of this project does not indicate that there is any legal basis for taking any National Guard units away from states without their consent and resubordinating them under the direct authority of the National Guard Bureau.

---

[23] Policy-level interviews with members of the office of the Assistant Secretary of Defense for Reserve Affairs January 25, 2011.

[24] This issue was reported by the Rhode Island 19th SFG contingent. During the project's February 22, 2011, briefing at USASOC; however, we were told that a recent review of ARNG Special Forces equipment requirements noted no such delivery problems.

[25] A suggestion from a National Guard general officer familiar with Special Forces equipping issues, December 6, 2010.

[26] Texas members of the 19th Special Forces Group during a site visit, January 18, 2011.

Other recommendations may be more tractable.[27] Many of these called for closer cooperation (especially at mobilization and afterward) between the ARNG Special Forces on one hand and USASOC/the AC Special Forces on the other. These recommendations included:

- Build USASOC-owned mobilization facilities, co-located with the directed training alignment (DTA) AC units so that, when mobilized, the ARNG units could "fall in" on their AC counterparts for better pre-deployment coordination and training.
- Turn DTA into "adoption" so that, once mobilized, the ARNG Special Forces effectively integrate into an AC unit.
- Ensure that post-mobilization training is conducted with the AC counterpart unit.

Other recommendations emphasized pre-mobilization improvements, including:

- Create bonuses to attract AC members who leave active duty prior to retirement. Advocates believe such a bonus would "fix the DMOSQ problem" in the ARNG Special Forces.
- Synchronize force modernization and equipping upgrades to the Playbook. Doing so would ensure that modernization and upgrades would be in place in time for programmed mobilizations.
- Provide AC advisors at company level. Units that had had such advisors found them invaluable.
- Return to the practice of issuing mission letters. Such letters would remove much of the uncertainty and ambiguity that persists in outlying units about their mission, mission-essential tasks, and expectations about future employment.
- Make greater use of Special Operations Detachments (SODs) to provide administrative and logistical assistance to outlying companies.
- Replace the Playbook with a real, detailed ARFORGEN process. The expectation here is that in doing so, the ARNG Special Forces would enjoy fuller transparency into future plans, especially deployments, and that transparency would lead to equitable treatment (e.g., regular deployments).
- Seek involuntary mobilization authority for non-named operations. Doing so would improve access to ARNG Special Forces for theater security cooperation activities as well as for non-named military operations.
- Provide senior ARNG Special Forces personnel (lieutenant colonels and above) need with advanced PME to develop their skills.

Niches for the ARNG Special Forces

The discussion of niches took three basic forms. The first emphasized the level of units within the ARNG Special Forces that were, in the judgment of the respondent, suitable for deployment. The second considered the types of operations, or tasks, or missions for which the ARNG seemed well suited. A third constituency rejected the notion of niches altogether, emphasizing

---

[27] All of the recommendations appearing in this section were collected during site visit discussions with members of the 19th and 20th Special Forces Groups, various dates, December 2010 through February 2011.

the fact that the ARNG Special Forces undergo the same training as the AC, and that therefore, they are just as capable as the AC and don't need niches, which they suspect would contain only those missions the AC prefers not to perform.

Suitable Units

The interview respondents' judgments, like the survey results, indicated that ODAs and ODBs were the most suitable ARNG Special Forces units for deployment. Several also indicated that individual mobilization augmentees (IMAs) were, on an individual basis, also fit for deployment.

Suitable Missions and Tasks

These discussions generated a list of specific missions that might be niches for ARNG Special Forces. The list includes:

- Foreign Internal Defense.
- The Afghan village security operations program and embedded tactical trainers.
- Super JCETs (60 days) that would "get us into AFRICOM without counting against mobilization time."
- Horn of Africa–like missions: small scale, remote, unattractive to AC Special Forces.
- Border security in the NORTHCOM area of responsibility.
- Theater security cooperation activities.

Chapter Conclusions

This chapter has in part raised issues that bear on the capabilities and potential contributions of ARNG Special Forces, and in part identified the Special Forces community views of at least some of the menu items that USASOC might adopt as part of its efforts to enhance ARNG Special Forces contributions. In particular, the chapter has identified:

- Opportunities to enhance cooperation and coordination between the ARNG and the AC Special Forces, most of which would require initiative on the part of USASOC or USASFC in order to realize them.
- Constraints that limit the value and utility of ARNG Special Forces.
- Options to enhance their performance, including some policies and actions affecting pre-mobilization operations, and others intended to influence post-mobilization training and operations.
- Niches where, at least in the judgment of some of the study's respondents, ARNG Special Forces should be expected to perform well.

Chapter Four draws upon Chapters Two and Three to provide USASOC with a menu of options to (1) populate the ARNG Special Forces problem space, and (2) create a blueprint to move the ARNG toward being a purpose-driven force.

USASOC's Menu of Options

The preceding two chapters were diagnostic and descriptive in nature. Chapter Two employed quantitative data provided by the National Guard Bureau, USASOC, and the 19th and 20th SFGs themselves to characterize their qualifications, readiness, and deployment history. The Special Forces survey provided additional data on respondents' years of service, individual qualifications, deployment history, willingness to make future deployments, views on unit capabilities, and characterizations of the respondents' civilian skills/careers.

Chapter Three provided generally qualitative data based upon policy-level interviews and site visit discussions that added nuance and richness to Chapter Two's conclusions. Chapter Three presents a consensus view of ARNG Special Forces and steps that could be taken to improve their contributions within U.S. Army Special Forces Command.

Chapter Four presents our analysis and recommendations in response to the individual conclusions at the ends of Chapters Two and Three.

Overall Conclusions and Recommendations

There are a few, but not many, statutory and policy issues that interfere with USASOC's ability to maximize the contributions of ARNG Special Forces or move them toward a "purpose-driven force."

USASOC does not enjoy complete freedom of movement. It still needs governors' consent in order to move ARNG units. There is no provision to remove ARNG units from their subordination to the adjutants general and place them directly under the National Guard Bureau, as some ARNG Special Forces personnel advocated. The high value that states place on having Special Forces as part of their Army National Guard almost surely means an impossible fight if USASOC were to try to move the Special Forces to the U.S. Army Reserve, which lacks the political clout of the National Guard and the state governors.

Nevertheless, adjutants general have a history of managing their units cooperatively and arranging bilateral tradeoffs when a state either lacks the resources to support a Special Forces unit or for other reasons (e.g., disaster response plan requirements) comes to prefer a different sort of unit. USASOC should observe this process and learn how to participate—at least from the sidelines—to ensure that ARNG Special Forces units always enjoy the full support of the states in which they are stationed and to support moves when doing so seems appropriate in terms of ensuring unit readiness.

In particular, DoDI 1235.10 (Enclosure 2) casts the National Guard as an asset of last resort, and although this characterization is apparently not treated as authoritative, it might usefully be rewritten to reflect the type and intensity of integration DoD expects from the AC and the ARNG.

From our inquiries among senior OSD officials, we conclude that the current edition of this DoD Instruction is not interfering with USASOC's ability to gain access to ARNG Special Forces. Nevertheless, DoD Instructions can be powerful instruments; if USASOC wants a particular relationship with its ARNG units or wants to specify certain characteristics of that relationship, this DoDI might be rewritten to characterize the desired relationship. USASOC could, through its chain of command, petition OSD for assistance in this regard.

The current chain of command provides limited functionality before mobilization and has prompted widespread, though not universal, suspicions of unfairness and favoritism.

There is no apparent remedy to the issues of complexity and limitations associated with the current National Guard chain of command. If USASOC wants to improve unity of effort across the ARNG portion of its forces, reduce uncertainty about future assignments, and dispel destructive rumors and suspicions of favoritism, it must look for other mechanisms beyond the chain of command for assistance. Periodic conferences could be one tool. A secure website to support cooperative planning, programming, and budgeting activities could be another. Mission letters and revitalized DTA relationships could be still others.

Supply of ARNG Special Forces is ample in terms of manpower, but below goal in terms of duty MOS qualification.

If USASOC determines it wants to increase the MOS qualification rate among the current manpower within the 19th and 20th SFGs, it must increase the quotas these groups receive to attend the Special Forces Qualification Course. The quotas by themselves will not be sufficient; it will be important to coordinate the number and timing of the quotas carefully with each SFG, and to ensure that the requisite pay and allowances have been programmed to support the candidates who attend the training.

If USASOC will consider the notion of paying bonuses to soldiers leaving the AC short of retirement to induce them to integrate into the ARNG SFGs, this might be another remedy to poor MOS qualification rates. If we assume a total population of 2,000 MOS 18-series positions within the ARNG, and assume that 70 percent of them are MOS qualified, that would leave 600 unfilled positions to be filled by soldiers leaving active duty and accepting bonuses of $50,000 apiece (the amount quoted during our site visit) to join the ARNG SFGs: a total bill of $30 million. To reach 85 percent MOS qualified, the bill is $15 million.

Despite the limitations in terms of duty MOS qualification, the ARNG Special Forces contain additional capacity and will to undertake future deployments.

Additional will to deploy, discovered in the responses to the Special Forces survey, becomes additional capacity to deploy when it either leads to individual augmentees or supports cross-leveling to render an ODA or ODB ready to deploy. USASOC might take advantage of this will to deploy in the future by operating a web-based bulletin board to solicit individual augmentees or to announce updates to the Playbook which show the next series of units to be mobilized for deployment.

In addition, USASOC could work with the National Guard Bureau and OSD to gain access to ARNG Special Forces for theater security cooperation activities and non-named operations: logical activities for taking advantage of the additional will to deploy.

The ARNG Special Forces generally have fewer personnel qualified in additional skills (e.g., military free fall, dive) than the AC, although the ARNG units have more snipers and their members typically are somewhat older and have more years of service than their AC counterparts.

The current demand signal includes many commitments where advanced skills are less critical to mission success. Examples include theater security cooperation activities, some operations in the Horn of Africa, and often, some FID missions. These could be good uses of ARNG Special Forces, where their slightly older personnel and rich experiences from civilian life might make them especially effective.

Language skills, a challenge throughout the force, do not appear to be a principal asset of the ARNG Special Forces.

Units with limited salient language capabilities (both AC and ARNG) could nevertheless be usefully employed in activities where language qualifications are least critical. Oftentimes, this would mean theater security cooperation activities, JCETs, and some FID missions.

Deployments are not only important to accomplish the mission, but are also essential for growing and maintaining competence within ARNG Special Forces and for winning the confidence of the AC.

We conclude that, at least under the prevailing conditions, it will be important to deploy all ARNG Special Forces units at some reasonable frequency in order to maintain their essential operational competence, and to ensure that they continue to enjoy some level of confidence with the units they work with. Arriving at a reasonable frequency for their deployment might derive from consideration of several factors: a desire to ease the BOG:dwell ratio for AC soldiers; the opportunity to exploit some ARNG attribute or proven capacity residing within a given unit; or the imperative to refresh operations skills before they decay beyond some threshold that USASOC is unwilling to tolerate.

The prevailing view in the AC of the ARNG Special Forces is guarded, but far from dismissive. Conditional consensus appears to exist for the premise that there are tasks and circumstances suitable for the ARNG, especially at ODA, ODB, and CS/CSS unit levels.

Many of the Guardsmen we encountered during site visits echoed these views. The consensus, insofar as it exists, is that at ODA, ODB, and CS/CSS unit level, ARNG Special Forces are capable units for FID, UW, JCETs, and other theater security cooperation activities. We have found no compelling evidence to refute this consensus.

There is also consensus that the senior leaders in the ARNG, typically lieutenant colonels and above, have limited opportunities for professional development. Therefore, insofar as USASOC wants to develop these ARNG officers, it must find ways to employ them in jobs that will support growing their expertise and capabilities. One option might be to create nominative assignments to bring a promising ARNG officer along on a deployment as a member of an AOB, the SOTF, or the JSOTF staff. Whatever option USASOC may choose to enhance professional development, it will be important to emphasize the officer's exposure to useful experiences. It will be equally important for USASOC to ensure that such officers are not simply brought along and then stuck in a do-nothing role.

Both women in the ARNG Special Forces Groups and civilian skills of the members, while important, do not appear to hold the additional potential benefits that USASOC may have hoped for.

If USASOC believes it needs women for specific tasks, it should seek them from within the general purpose forces. Special Forces has done this historically, deploying ODAs with medical teams featuring female nurses and physicians. Not all specialists must necessarily come from within the special operations community to be effective partners, although many tasks will require that newcomers undergo familiarization training and preparation for their specific roles.

If USASOC believes it remains important to tap specific civilian skills for some deployments, the bulletin board recommendation described earlier in this chapter may be indicated.[1] Advertise for the skills desired, and encourage soldiers with those skills to volunteer. In some instances where the skills in question may be relatively rare, USASOC may consider bonuses or other incentives to attract qualified personnel.

Demand for Special Forces is high, though not at its apogee, and unstable. It appears that ARNG Special Forces can play a useful role in satisfying demand, especially demand that takes the form of scheduled events: programmed rotations within the Playbook and theater security cooperation activities.

The ARNG Special Forces record of deployments is documented at USASFC. If USASOC wants to increase the ARNG Special Forces ability to satisfy demand, it could do at least three things.

Many of those interviewed also believe that theater security cooperation events will likely have increased lengths. Some have called these super-JCETs. With increased length of events, USASOC and USASFC will have to ensure funding to cover ARNG-unique costs, namely pay and allowance.

Second, USASOC could give the National Guard units greater priority in scheduling within the Playbook to maximize the quality of fit between their skills and the missions for which they are mobilized and deployed. We understand that there are practical constraints that limit USASOC's ability to preserve this priority, but wherever possible, doing so would ensure the best match of supply and demand where the ARNG Special Forces are concerned.

Third, it could include the leadership from the two ARNG SFGs in a force generation conference to update the Playbook periodically. Including the 19th and 20th SFG leadership would be useful in aligning their subordinate units against future requirements for Special

---

[1] An online job board would need to be on an unclassified system to enable ARNG Special Forces members to gain access to the information. Many may not have regular access to classified systems.

Forces, and might make an important improvement in the quality of the "fit" between demand and supply.

Opportunities exist to enhance cooperation and coordination between the ARNG and the AC Special Forces, most of which would require initiative on the part of USASOC or USASFC in order to realize them.

We believe USASOC has several options for enhancing cooperation and coordination between the AC and ARNG. Conferences could accomplish much of the additional interaction. For example, a POM planning conference could ensure that both components share the same expectations in terms of training, equipping, pay and allowances, and other issues, and build their POMs to reflect the funding needed to support these various activities. We imagine such a conference as a modest endeavor, with the 19th and 20th SFG commanders and their financial management staff members meeting with the appropriate USASOC staff and National Guard advisor to accomplish the necessary coordination.

Reviving the old SOF/Adjutant General conference could provide cooperation and coordination over broader policy issues. Guardsmen involved in past conferences noted that they did not have much influence on the agenda. If USASOC decides to reinvigorate the annual conferences, it should take steps to ensure that the National Guard participants can play a constructive role in shaping the agenda.

Other mechanisms could help USASOC enhance cooperation and coordination. A return to robust DTA relationships could be one such option. The directed training alignment, however, to be useful, must endure so that both the National Guard and AC participants view it as a habitual association with real substance and depth.

Mission letters could also be valuable tools. We understand that USSOCOM does not look favorably on mission letters, but in the unique circumstances that dictate the AC/National Guard relationship, we believe a return to mission letters could be a very valuable tool for focusing ODA-level training, country focus, and preparations for future mobilizations.

Finally, a SIPRNET-based website devoted to coordinating instructions between USASOC and the 19th and 20th SFGs could keep both USASOC and the groups apprised of the groups' current status, coming events, looming issues requiring cooperative planning and coordinated responses, changes to the Playbook, and similar critical information.

Constraints exist that limit the value and utility of ARNG Special Forces.

This report has documented many of the constraints that limit the value and utility of ARNG Special Forces. A point for the AC to consider, however, is that Special Forces generally and routinely work with less capable partners, including commandos from security cooperation partners, or tribal militias scattered through the hinterlands of the developing world. Special Forces routinely figure out ways to limit and manage the risk inherent in operating with less capable partners; surely they can—and are—figuring out how to employ ARNG Special Forces to maximum effect without incurring unacceptable risk. The niches discussed in Chapter Three suggest such applications.

Options exist to enhance their performance, including some policies and actions affecting pre-mobilization operations, and others intended to influence post-mobilization training and operations.

We conclude that predictability, reasonable lead time, and mission letters are key ingredients for enhancing the performance of the ARNG Special Forces. Mission letters could be the instrument that focuses individual ODAs on the skills and missions they must perfect in order to be successful in their next mobilization and deployment. The mission letter in this sense is the key item assuring predictability—that when the mobilization order arrives, it will result in the team being deployed to perform the tasks documented in its mission letter.

Lead time becomes crucial for managing change. ARNG units need more of it in order to respond to change because they have relatively less training time available than their AC counterparts. Thus, if USASOC wants to enhance the performance of these units, it should provide good mission letters and do everything possible to preserve the taskings within those letters. When the inevitable happens and a team's letter must be modified, that modification should be made as early as practical so that the team has maximum lead time in which to adjust.

USASOC could take other, more vigorous steps beyond mission letters if it believes the demand for Special Forces is likely to remain high, and therefore, that continued access to the ARNG SFGs will be desirable. It could support construction of mobilization facilities at the DTA AC unit's location and require the AC DTA unit to become more involved in assisting the ARNG unit with its post-mobilization training: scheduling ranges, planning joint training, and so forth. Even if the military construction costs for new mobilization sites appear prohibitive to USASOC, it could still require USASFC to provide post-mobilization training support from the AC DTA unit to maximize the training content for the mobilizing unit.

Toward Purpose-Driven ARNG Special Forces

So, based on this project's research, what might purpose-driven ARNG Special Forces look like, and how would they operate? We offer some thoughts couched in terms of organization, training, equipping, manning, and employment.

The purpose-driven ARNG Special Forces unit's organization is requirements-driven. That is, it morphs to accomplish the mission, just as its AC forebears have done over the decades. We expect that the preponderance of deployed ARNG units will be ODA- and ODB-level formations, because that seems to reflect the prevailing consensus on their utility and competence. Even so, these organizational types could be modified to complement their DTA AC unit and the missions it is preparing to conduct.

Pre-mobilization individual training will emphasize duty MOS qualification: get the maximum number of soldiers through the Special Forces Qualification Course. All training planning and programming would make 85 percent MOS qualification the goal. Unit training would focus on the mission-essential task list agreed to with the DTA AC unit and USASFC.

Pre-mobilization equipping would be just in time and requirements-driven. The unit's new equipment reception and training plan would be synchronized through an ARFORGEN-like process to get it to the end users in time for their final, pre-mobilization training.

Manning would emphasize team integrity and feature AC advisors/mentors at ODA level. Where the ARNG Special Forces unit could not achieve critical mass (e.g., generate at least nine qualified soldiers per ODA), the personnel from ODA in question would be used as individual fillers on other teams.

Employment of ARNG Special Forces would be motivated by three factors. First, they would be employed to moderate the demand for AC Special Forces. Second, ARNG Special Forces would be employed for tasks, missions, and assignments (e.g., JCETs, theater security cooperation missions, some UW and FID assignments) that free AC Special Forces to perform more demanding tasks. In some instances ARNG Special Forces would be employed in lieu of AC Special Forces (e.g., for JCETs, theater security cooperation). In other cases, ARNG Special Forces would be employed as an integral part of their DTA AC unit (e.g., in a SOTF or CJSOTF).

Implementation: Making the ARNG Special Forces Purpose-Driven

Implementation could follow a four-step process described below to embrace the actions identified in this report.

The colored numbers to the left of each listing in Figure 4.1 suggest an order of implementation, based upon a logic reflecting the authorities available to USASOC and the costs of implementing each option. Figure 4.2 illustrates. According to Figure 4.2, USASOC should implement those actions whose costs are low, and that can be done on the organization's own authority.

These include employing ARNG Special Forces for tasks including theater security cooperation (TSC), JCET, FID, UW, Combined Joint Task Force–Horn of Africa Building Partner Capacity operations, and extended training operations. They also include employing ARNG Special Forces to ease the operations tempo for the active component units. These recommendations also consider the appropriate units for employment, emphasizing ODAs, ODBs, and SOTFs as the most appropriate size formations for ARNG Special Forces to command and

Figure 4.1
Recommendations

① • Employ ARNG Special Forces for recommended tasks (TSC, JCET, FID, UW, Horn of Africa–like, Afghan village security, etc.)
  • Deploy to manage active component OPTEMPO
  • Emphasize employment of ODA, ODB, and SOTF
  • Operate Internet site to solicit volunteers based on their civilian skills
  • Renew use of mission letters
  • Guaranteed deployments to maintain skills

② • Regular Army advisors at SF company level

③ • More Special Forces Qualification Course quotas and support
  • Extended Playbook
  • Revitalize directed training alignment (DTA) relationships
  • Sponsor more coordination and planning conferences
  • Sponsor nominative assignments for promising senior ARNG Special Forces officers

④ • Seek authority for access to ARNG Special Forces for non-named operations
  • Create mobilization sites at DTA active component home station
  • Create proportionate force structure to facilitate rotations

RAND *TR1199-4.1*

Figure 4.2
Implementing the Study's Recommendations

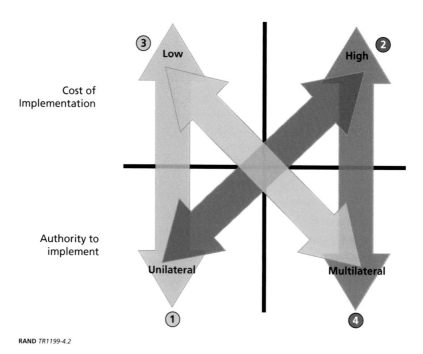

control. When USASOC seeks to tap individual skills, it could operate an Internet website to solicit volunteers based upon their civilian skills. Finally, the inexpensive, unilateral recommendations advocate for the renewed use of mission letters to specify mission-essential tasks for each ODA, and to ensure that all ARNG Special Forces undergo some minimum number of operational deployments to maintain their skills and the confidence of their AC counterparts, with whom they typically operate when deployed.

The second class of recommendations—those that are unilateral but expensive—contains a single recommendation. USASOC should return to the practice of assigning Regular Army advisors at Special Forces company level. Virtually everyone we encountered had a positive view of this practice and saw it as a very effective way to transmit recent operational experience and tactics, techniques, and procedures (TTP) into the ARNG.

The third category of recommendations includes those actions that are relatively inexpensive but require multilateral agreement and coordination. There are five such actions. The first of these is for USASOC to send the ARNG more Special Forces Qualification Course (SFQC) quotas and to task U.S. Special Forces Command (USASFC) to work with the ARNG units and state adjutants general to prepare Guardsmen candidates, support them and their families during the course, and produce a higher graduation rate. The second recommendation in this category is for USASOC to extend the Playbook and share its contents earlier so that ARNG units have better insight into when they will be mobilized next, where they are likely to be deployed, and what missions they are likely to perform. Third, USASOC should revitalize directed training alignments between the AC and ARNG Special Forces. Ideally, the ARNG mobilization sites should be co-located with their DTA AC unit and they should deploy together. Falling short of that, the Regular Army company-level advisors should come from the DTA AC unit, and the DTA units should coordinate all collective training with the

ARNG units aligned with them. Fourth, USASOC and USASFC should sponsor more conferences to conduct planning and coordination with the ARNG units. For example, force generation conferences and the process of building the Playbook should involve the ARNG SFG commanders. Finally, in order to enhance senior leader (lieutenant colonel and above) capabilities within the ARNG Special Forces, USASOC/USASFC should sponsor nominative assignments that would afford promising ARNG officers the opportunities to deploy in AC staff and command jobs and to gain experience under the direct supervision of AC seasoned experts.

The final category of recommendations—both expensive and requiring multilateral coordination and agreement—includes three actions. USASOC should seek authority to access its ARNG Special Forces involuntarily for non-named operations. Such authority would make it much easier to employ the ARNG to manage AC operations tempo. As previously noted, there are broader efforts ongoing to make access to ARNG units easier for non-named operations. Second, USASOC/USASFC should create mobilization sites at the DTA home stations so that the ARNG Special Forces would mobilize and fall in on their AC counterparts. Third, insofar as USASOC must sustain a smooth rotation of forces in overseas contingency operations and direct interchangeability of units is desirable, USASOC should create proportionate units in the ARNG.

Brief History of Reserve Component Special Forces

Reserve component Special Forces units began to appear in the late 1950s. Some of the first units included the 300th, 301st, 302nd, and 303rd FD Operational Detachments. At the height, RC SF included 11 different SFGs. Seven SFGs were in the reserves: the 2nd, 9th, 11th, 12th, 13th, 17th, and 24th SFGs; while the ARNG had four: the 16th, 19th, 20th, and 21st SFGs. Most of the SFGs were not fully filled out, with some containing only a company. Many of the SFGs began be deactivated in 1966.

The RC SFGs consolidated into two SFGs in the reserves (11th and 12th) and two in the ARNG (19th and 20th) by the 1990s. Operation DESERT SHIELD/DESERT STORM saw the first activation of a RC SFG, the 20th SFG in the ARNG. The conflict ended prior to the 20th's deployment, but some of its members participated in Operation PROVIDE COMFORT.

With the end of the Cold War, DoD began searching for ways to decrease costs. In November 1990, it developed guidance to deactivate three ARNG Special Forces battalions and three reserve Special Forces battalions.[1] Shortly afterward, DoD rescinded the inactivation of the three reserve Special Forces battalions until a joint mission analysis was conducted.[2]

The U.S. Congress also got involved. The 1992 Defense Appropriations Act prevented the conversion of the ARNG Special Forces missions to the active component, while the 1993 DoD Appropriations Act Report went further and rejected any possibility to expand the active component to replace the ARNG SFGs.[3] The joint mission analysis validated the need to deactivate six battalions, three from the reserves and three from the ARNG.[4] To the surprise of many, on September 15, 1995, both reserve SFGs (11th and 12th) were deactivated.

---

[1] Government Accountability Office, *Special Operations Forces: Force Structure and Readiness Issues*, March 1994, NSIAD-94-105, p. 29.

[2] Government Accountability Office, *Special Operations Forces: Force Structure and Readiness Issues*, March 1994, NSIAD-94-105, p. 29.

[3] Government Accountability Office, *Special Operations Forces: Force Structure and Readiness Issues*, March 1994, NSIAD-94-105, p. 29.

[4] Government Accountability Office, *Special Operations Forces: Force Structure and Readiness Issues*, March 1994, NSIAD-94-105, p. 29.

Annotated Bibliography of Collected Sources

This appendix summarizes the applicable statutory and regulatory requirements that dictate how ARNG Special Forces will be trained, equipped, manned, and utilized. It is meant mainly as a quick reference on relevant U.S. Code and policies. It provides a one-line summary followed by the major implications for ARNG Special Forces and USASOC.

Table B.1
Annotated Bibliography of Collected Sources

United States Code, Title 10—*Armed Forces,* February 1, 2010 ed. Accessed at http://www.law.cornell.edu/uscode/, October 5, 2010.	
SUMMARY:	This section of U.S. Code provides the statutory requirements and restrictions for the Federal Armed Services.
MAJOR IMPLICATIONS:	• Describes how ARNG personnel and units will be handled when in federal service • Describes general authority of combatant commanders, including Commander, USSOCOM • Identifies the general types of special operations • Describes budget support to reserve elements
United States Code, Title 32—*National Guard,* February 1, 2010 ed. Accessed at http://www.law.cornell.edu/uscode/, October 5, 2010.	
SUMMARY:	This section of U.S. Code describes the organization, relationships, and roles for the state National Guards.
MAJOR IMPLICATIONS:	• Except as provided by USC, the organization of ARNG and composition of ARNG units will be the same as those prescribed for the Army • Prohibits movement of units located in particular states without that state governor's consent • Describes the role of ARNG in drug-interdiction and counter-drug activities • Describes federal financial assistance to and equipment procurement for the ARNG • Regulates employment of technicians in the ARNG • Describes homeland defense roles for the ARNG
Department of Defense, *DoD Directive 1200.17: Managing the Reserve Components as an Operational Force.* Washington, D.C.: Government Printing Office, 2008.	
SUMMARY:	This directive provides general guidance from the Secretary of Defense to establish the overarching set of principles and policies to promote and support the management of the Reserve Components (RCs) as an operational force.
MAJOR IMPLICATIONS:	• Identifies applicable U.S. Code provisions • Describes role of RC forces in homeland security • Identifies the basic value of the RC forces as deployable forces • Encourages voluntary service among the RC forces to meet requirements • Requires RC resourcing plans to provide visibility of resources

Department of the Army, *AR 10-87: Army Commands, Army Service Component Commands, and Direct Reporting Units.* Washington, D.C.: Government Printing Office, 2007.	
SUMMARY:	This regulation defines the chain of command and relationship of Army commands and units.
MAJOR IMPLICATIONS:	• Describes the roles and responsibilities of USASOC as an Army Service Component Command • Establishes USASOC's responsibility to oversee and evaluate ARNG Special Forces in CONUS and ensure that these units are organized to accomplish SOF missions and support appropriate OPLANs. • Specifies the relationships that USASOC command and staff will maintain with National Guard Bureau and appropriate state's adjutants general
Department of the Army, *AR 11-30: Army WARTRACE Program.* Washington, D.C.: Government Printing Office, 1995.	
SUMMARY:	This regulation dictates the deliberate alignment of Army forces (active component as well as reserve component) under a single commander for wartime planning to achieve national strategic goals under the WARTRACE program.
MAJOR IMPLICATIONS:	• Defines the Chief, National Guard Bureau's responsibility to inform WARTRACE alignments between ARNG units and FORSCOM (general purpose forces) and USASOC • Directs coordination of FORSCOM and USASOC in WARTRACE planning • Directs USASOC to coordinate directly with NGB to implement ARSOF WARTRACE plans for RC ARSOF units • Requires peacetime commanders to provide resources and ensure all training for wartime mission readiness
Army National Guard, *NGR 350-1: Army National Guard Training.* Washington, D.C.: Government Printing Office, 2009.	
SUMMARY:	Describes the requirements for ARNG training, assembly, and planning.
MAJOR IMPLICATIONS:	• Identifies required amount of training time all ARNG personnel must participate in annually • Defines USASOC's roles and responsibilities for training ARNG SOF forces • Provides provisions for annual training and year-round annual training for ARNG personnel • Provides provision for ARNG SOF deployment for training • Identifies major SOF units in the ARNG
Army National Guard, *NGR 614-2: Army National Guard Airborne and Special Forces Units.* Washington, D.C.: Government Printing Office, 1979.	
SUMMARY:	This regulation prescribes procedures for selecting and assigning ARNG personnel to airborne and Special Forces units, including training requirements and security provisions.
MAJOR IMPLICATIONS:	• Indicates that instruction to ARNG to undergo airborne and Special Forces training will come from HQDA • Requires ARNG personnel assigned to airborne or Special Forces slots to attend appropriate training within 12 months
United States Army Special Operations Command, *USASOC Reg 350-1: Army Special Forces (ARSOF) Active Component and Reserve Component Training.* Washington, D.C.: Government Printing Office, 2005.	
SUMMARY:	This USASOC document provides the overarching guidance to all USASOC active and reserve component elements concerning training requirements.
MAJOR IMPLICATIONS:	• Identifies specific responsibilities of USASOC commander and staff concerning the ARNG SOF units • Delineates the responsibilities of the state adjutants general to ensure ARNG SOF unit readiness • Provides specific additional and supplemental training objectives for RC elements • Identifies specific training deployment requirements for RC ARSOF units • Identifies the recommended frequencies for AC and RC USASOC elements at each echelon

Department of the Army, *AR 350-9: Overseas Deployment Training*. Washington, D.C.: Government Printing Office, 2004.

SUMMARY:	Describes restrictions and responsibilities on combatant commands for planning, coordinating, and conducting overseas deployment training (ODT).
MAJOR IMPLICATIONS:	• Indicates Director, ARNG's role, as a force provider, in planning and execution of ODTs • Dictates Commander, USASOC's role and responsibilities for coordinating ODT • Explains requirements for overseas coordination conference in preparation for ODT • Identifies how in-country costs will be accounted for

Department of Defense, *DoD Directive 1225.6: Equipping the Reserve Forces*. Washington, D.C.: Government Printing Office, 2005.

SUMMARY:	Directive describes the responsibilities, standards, and processes for equipping the reserve component based on mission requirements and unit composition.
MAJOR IMPLICATIONS:	• Requirements for equipping will be derived by applying the same methodology as AC units that have the same mission requirements • Requires Secretary of Defense approval for withdrawals, diversions, or reductions of RC component equipment • Allows equipment to be drawn from Ready Reserve units to meet the requirements of mobilized Ready Reserve units or AC units supporting the same mission

Department of Defense, *DoD Directive 1200.16: Contracted Civilian-Acquired Training (CCAT) for Reserve Components*. Washington, D.C.: Government Printing Office, 2004.

SUMMARY:	This directive provides guidance for use of civilian contracted training to meet training requirements that cannot be met through standard training resources.

MAJOR IMPLICATIONS:	• Contracting is authorized to meet RC training requirements if the military service school system is:
	   ○ Unavailable
	   ○ Not practicable due to travel costs

Department of Defense, *DoD Directive 1235.10: Activation, Mobilization, and Demobilization of the Ready Reserve*. Washington, D.C.: Government Printing Office, 2008.

SUMMARY:	This directive describes the procedure and responsibilities for activating, mobilizing, and demobilizing RC from Title 10 federal service.

MAJOR IMPLICATIONS:	• SoA assesses a unit's readiness for mobilization against the unit's assigned METL, based on:
	   ○ Evaluation of the unit's ability to perform tasks to prescribed standards
	   ○ Under conditions expected in theater operations where unit would be deployed

Department of Defense, *DoD Directive 1250.01: National Committee for Employer Support of the Guard and Reserve*. Washington, D.C.: Government Printing Office, 2007.

SUMMARY:	This regulation outlines the composition and role of the committee that promotes public and private understanding of the ARNG to gain employer and public support.
MAJOR IMPLICATIONS:	• Defines Committee to include representatives at the state and national headquarters levels

Department of Defense, *DoD Directive 5105.77: National Guard Bureau*. Washington, D.C.: Government Printing Office, 2008.

SUMMARY:	Describes the composition, roles, and authorities of the National Guard Bureau for all National Guard RC elements.
MAJOR IMPLICATIONS:	• Designates NGB as the channel of communication for all matters pertaining to the National Guard between the Dept. of the Army and the states • Requires SoA to be informed of all significant matters pertaining to the Army • Implements DoD and Army guidance on structure, strength, authorizations, and other resources of the ARNG • Prescribes the training requirements and allocation of federal funds for training the ARNG

Department of Defense, *DoD Directive 1235.12: Accessing the Reserve Component (RC)*. Washington, D.C.: Government Printing Office, 2010.	
SUMMARY:	This regulation prescribes procedures for ordering units and individual members of the RC to active duty as an operational force.

MAJOR IMPLICATIONS:	• Preparing RC for contingency operations requires a two-step process:
	  ○ Activating by DoD by ordering RC to active duty
	  ○ Mobilizing RC by preparing them for operational missions

Department of Defense, *DoD Instruction 3305.06: SOF Foreign Language Policy*. Washington, D.C.: Government Printing Office, 2008.	
SUMMARY:	This instruction provides guidance for development, sustainment, and management of individual foreign language skills for effective conduct of SOF core missions and activities.
MAJOR IMPLICATIONS:	• Provides CDR USSOCOM authority to determine SOF organization language requirements • States foreign language skills will be an element of readiness for SOF units that have requirements for language-skilled SOF • States CDR USSOCOM will identify foreign language requirements to support operational needs of SOCOM and geographic combatant command SOF operations

Chairman, Joint Chiefs of Staff, *CJCS Instruction 3126.01: Language and Regional Expertise Planning*. Washington, D.C.: Government Printing Office, 2008.	
SUMMARY:	Instruction provides guidance to plan for regional demand for language and other expertise to support impending operations.
MAJOR IMPLICATIONS:	• States USSOCOM will consolidate, track, and manage all SOF foreign language support requirements for geographic COCOMs • States CDR USSOCOM has the authority and responsibility to train and organize SOF to support geographic COCOM and select USSOCOM directed missions

Department of Defense, *DoD Instruction 1235.11: Movement of Individual Mobilization Augmentees*. Washington, D.C.: Government Printing Office, 2007.	
SUMMARY:	Instruction provides guidance for use of RC individuals to serve on active duty (Title 10) support AC units and operations.
MAJOR IMPLICATIONS:	• Allows IMAs to fill billets in AC units designated for fill by RC members • Gives organizations with high-priority mobilization missions priority for IMAs • States IMAs must receive appropriate training for the AC billet they will fill

Department of Defense, *DoD Instruction 1205.18: Full-Time Support to the Reserve Component*. Washington, D.C.: Government Printing Office, 2007.	
SUMMARY:	Instruction provides guidance for use of full-time service members (FTS) to assist RC unit organization, administration, recruitment, instruction, training, maintenance, and supply support.
MAJOR IMPLICATIONS:	• States FTS will be assigned based on their military grade and skill codes • Ensures an FTS force will be established to be capable of ensuring accomplishment of RC readiness goals • States MTs will be used to maximize readiness • Priority resourcing will be given to high-priority and early deploying units.

Commission on the National Guard and Reserves, *Final Report to Congress*. Washington, D.C.: Government Printing Office, 2008.	
SUMMARY:	This report provides an extensive list of recommendations for improving the preparedness and effectiveness of reserve components as an employable force.
MAJOR IMPLICATIONS:	• Provides 32 recommendations for improving manning, equipping, training, and employment of the RC that likely have implications for USASOC's role in training and equipping the ARNG Special Forces groups

Department of Defense, *DoD Directive 5100.3: Support of the Headquarters of the Combatant and Subordinate Commands*. Washington, D.C.: Government Printing Office, 2007.	
SUMMARY:	This directive generally delineates the responsibilities of USSOCOM as a combatant command to administer and support special operations activities. This directive also defines special operations activities and special operations peculiar.
MAJOR IMPLICATIONS:	• States USSOCOM is responsible for administering all special operations and special operations peculiar activities (MFP-11)

Department of Defense, *Memorandum of Guidance for Developing and Implementing Special Operations Forces Program and Budget.* **Washington, D.C.: Government Printing Office, 1989.**

SUMMARY:	This memo provided initial guidance for USSOCOM for developing and implementing the Special Operations program and budget for MFP-11 and other programs to support SOF.
MAJOR IMPLICATIONS:	• Makes CDR USSOCOM responsible for programming and budgeting for SOF programs • Provides the military departments the responsibility for designing and approving force structure of SOF organizations

Department of Defense, *Memorandum for Managing Military Personnel Resources in the Defense Health Program and the Special Operations Command.* **Washington, D.C.: Government Printing Office, 2007.**

SUMMARY:	This memo outlines policies and procedures for programming and budgeting military personnel resources for USSOCOM.
MAJOR IMPLICATIONS:	• Allows CDR USSOCOM to transfer end strength within USSOCOM that does not result in a net change of military personnel end strength for the individual service components. • Provides technical instruction for calculating transfer price for military end strength transfers

United States Army, *Memorandum of Agreement (MOA) between the U.S. Army and the U.S. Special Operations Command.* **Washington, D.C.: Government Printing Office, 1993.**

SUMMARY:	This memo describes provisions and responsibilities as agreed upon by both the Army and SOCOM on matters pertaining to ARSOF planning, programming, mobilization, and resourcing.
MAJOR IMPLICATIONS:	• Makes SOCOM responsible for planning and programming for ARSOF • States that the Army will mobilize, deploy, redeploy, and demobilize ARSOF • States that the Army will administer both AC and RC ARSOF through USASOC

U.S. Army, *Annex F to Memorandum of Agreement (MOA) between the United States Army and the United States Special Operations Command (USSOCOM).* **Washington, D.C.: Government Printing Office, 1990.**

SUMMARY:	This Annex delineates transfer of ARSOF MFP-11 programs from the Army to SOCOM.
MAJOR IMPLICATIONS:	• Makes SOCOM responsible for all ARSOF MFP-11 programs • Provides diagrams describing budget formulation and funding distribution processes for ARSOF

Tabular Survey Results

This appendix provides summary statistics for selected questions from the ARNG Special Forces survey. The survey targeted the 19th and 20th Special Forces Groups because they contain the overwhelming majority of ARNG Special Forces personnel. The study sponsor also identified a subset of AC personnel to participate in the survey. The survey instrument was housed on a RAND website. We used the unit email alert system to make members aware of the survey and to solicit their participation. We conducted periodic follow-up inquiries with units to ensure that subordinate formations were aware of the survey and had the opportunity to respond to it.

Several steps were taken to clean and simplify the response data prior to generating statistics, including removing blank responses, identifying and removing duplicate responses, and recoding some fields to yield consistent responses for summaries. In total, 6 blank responses were removed, and another 23 were identified as duplicates and deleted (16 from a single respondent). In most cases, duplicate responses occurred when the respondent stopped and restarted the survey at some point in the process, and in these cases only the last response was retained. Starting from 383 entries, the data cleaning described above yielded 354 unique, non-null survey responses. Responses summarized in Tables C.2, C.3, C.6, C.7, and C.13 below reflect text responses that were standardized or categorized in order to generate summary statistics.

Table C.1
How old are you? (N=350)

Low	22 years
High	60 years
Median	38 years
Mean	37.75 years
Std. Dev.	7.67 years

Table C.2
What is your primary Military Occupational Specialty (MOS)? (N=350)

MOS	Freq.	Percent	MOS	Freq.	Percent
00Z	3	0.86%	35N	1	0.29%
11A	6	1.71%	35Y	1	0.29%
11B	3	0.86%	36B	1	0.29%
131A	1	0.29%	37A	1	0.29%
13A	3	0.86%	38A	1	0.29%
13F	2	0.57%	42A	6	1.71%
15A	1	0.29%	45B	1	0.29%
15E	2	0.57%	53A	1	0.29%
180A	29	8.29%	61N	1	0.29%
18A	44	12.57%	62A	2	0.57%
18B	25	7.14%	63A	1	0.29%
18C	21	6.00%	65B	1	0.29%
18D	19	5.43%	65D	1	0.29%
18E	23	6.57%	68J	1	0.29%
18F	19	5.43%	68W	1	0.29%
18Z	65	18.57%	74A	1	0.29%
21B	1	0.29%	74D	11	3.14%
250N	1	0.29%	79T	1	0.29%
254A	1	0.29%	89B	1	0.29%
25A	1	0.29%	90A	5	1.43%
25B	6	1.71%	91B	1	0.29%
25S	3	0.86%	91Z	1	0.29%
25V	1	0.29%	920A	1	0.29%
25W	2	0.57%	921A	1	0.29%
351L	1	0.29%	92R	3	0.86%
35D	2	0.57%	92W	1	0.29%
35F	6	1.71%	92Y	7	2.00%
35M	4	1.14%	No Response	4	

Table C.3
What is your rank? (N=348)

Code	Abbr.	Rank	Freq.	Percent
E-3	PFC	Private First Class	3	0.86%
E-4	SPC	Specialist	16	4.60%
E-5	SGT	Sergeant	21	6.03%
E-6	SSG	Staff Sergeant	55	15.80%
E-7	SFC	Sergeant First Class	73	20.98%
E-8	MSG	Master Sergeant	45	12.93%
E-8	1SG	First Sergeant	6	1.72%
E-9	SGM	Sergeant Major	18	5.17%
E-9	CSM	Command Sergeant Major	4	1.15%
W-1	WO1	Warrant Officer	7	2.01%
W-2	CW2	Chief Warrant Officer	17	4.89%
W-3	CW3	Chief Warrant Officer	7	2.01%
W-4	CW4	Chief Warrant Officer	2	0.57%
W-5	CW5	Chief Warrant Officer	2	0.57%
O-1	2LT	Second Lieutenant	1	0.29%
O-2	1LT	First Lieutenant	2	0.57%
O-3	CPT	Captain	34	9.77%
O-4	MAJ	Major	25	7.18%
O-5	LTC	Lieutenant Colonel	8	2.30%
O-6	COL	Colonel	2	0.57%
		No Response	6	

Table C.4
Are you Active Component, or National Guard? (N=347)

Response	Freq.	Percent
Active Component	80	23.05%
National Guard	267	76.95%
No Response	7	

Table C.5
If you are in the National Guard, were you ever a member of the Active Component? (N=267)

Response	Freq.	Percent
Yes	138	51.69%
No	129	48.31%

Table C.6
If you were a member of the Active Component, how much federal service do you have? (N=149)

Low	1 year
High	37 years
Median	8 years
Mean	8.75 years
Std. Dev.	5.56 years

Table C.7
If you are in the National Guard, what is your civilian occupation? (N=263)

Occupation	Freq,	Percent
Law Enforcement/Security	70	26.62%
Full Time National Guard	29	11.03%
Medical	21	7.98%
Non-Specific Small Business/Contractor	16	6.08%
Trades	15	5.70%
Student	14	5.32%
Defense Contractor	12	4.56%
Information Technology/Networks	11	4.18%
Fire Fighter	8	3.04%
Civil Service	7	2.66%
Corporate Management/Sales	6	2.28%
Attorney	5	1.90%
Intelligence	5	1.90%
Engineer	5	1.90%
Transportation (Air, Train, Boat)	4	1.52%
Teacher	2	0.76%
Communications/Public Affairs	2	0.76%
Medical Equipment	2	0.76%
Real Estate Management/Speculation	2	0.76%
Irregular Warfare Analysis	2	0.76%
Construction/Heavy Equipment	2	0.76%
Plant Operations	2	0.76%
Other	21	7.98%
No Response	4	

Table C.8
What type of unit are you in? (N=351)

Response	Freq.	Percent
Operational Detachment Alpha (ODA)	149	42.45%
Operational Detachment Bravo (ODB)	41	11.68%
Special Forces Battalion (SF BN)	53	15.10%
Special Forces Group Headquarters Company (SFG HHC)	63	17.95%
Support Company	37	10.54%
Other	8	2.28%
No Response	3	

Table C.9
Have you had any deployments since September 2001? (N=352)

Response	Freq.	Percent
Yes	308	87.50%
No	44	12.50%
No Response	2	

Table C.10
Are you an Active Component Special Forces member? (N=338)

Response	Freq.	Percent
Yes	71	21.01%
No	267	78.99%
No Response	16	

Table C.11
If you are an Active Component Special Forces member, on what basis were you involved with National Guard Special Forces? (N=50)

Response	Freq.	Percent
Active Component unit advisor	2	4.00%
Operational Control (OPCON)	9	18.00%
Tactical Control (TACON)	8	16.00%
Other	31	62.00%
No Response	21	

Table C.12
If you are an Active Component Special Forces member, were you a commander or staff officer/NCO at the time? (N=58)

Response	Freq.	Percent
Yes	26	44.83%
No	32	55.17%
No Response	13	

Table C.13
If you are an Active Component Special Forces member, how frequently did you observe or interact with your National Guard Special Forces? (N=57)

Response	Freq.	Percent
Daily	26	45.61%
Weekly	11	19.30%
Monthly	3	5.26%
Infrequently	2	3.51%
Never	15	26.32%
No Response	14	

Table C.14
Are you a member of the National Guard? (N=339)

Response	Freq.	Percent
Yes	261	76.99%
No	78	23.01%
No Response	15	

Table C.15
If you are a member of the National Guard, do you think your civilian skills could have been better utilized? (N=252)

Response	Freq.	Percent
Yes	183	72.62%
No	69	27.38%
No Response	9	

Table C.16
If you are a member of the National Guard and the circumstances were right, would you volunteer to deploy as an individual to fill a slot in an AC SF unit or to perform some other specified Special Forces function? (N=253)

Response	Freq.	Percent
Yes	227	89.72%
No	26	10.28%
No Response	8	

Table C.17
If you are a member of the National Guard, how frequently would you be prepared to deploy? (N=253)

Response	Freq.	Percent
1 year out of 5	21	8.30%
1 year out of 4	39	15.42%
1 year out of 3	181	71.54%
None of the above	12	4.74%
No Response	8	

Table C.18
If you are a member of the National Guard, regarding ODAs, when it comes to operational capabilities, are National Guard ODAs, when deployed, typically: (N=242)

Response	Freq.	Percent
About the same as their Active Component equivalent	64	26.45%
Somewhat limited in their capabilities relative to Active Component equivalent	23	9.50%
More capable at some tasks than Active Component equivalent	155	64.05%
No Response	19	

Table C.19
If you are a member of the National Guard, regarding companies and ODBs, in your experience, when it comes to operational capabilities, when deployed, are National Guard companies and OBDs typically: (N=238)

Response	Freq.	Percent
About the same as their Active Component equivalent	121	50.84%
Somewhat limited in their capabilities relative to Active Component equivalent	68	28.57%
More capable at some tasks than Active Component equivalent	49	20.59%
No Response	23	

Table C.20
If you are a member of the National Guard, regarding higher levels of command and control for special operations: in your judgment, when it comes to operational capabilities, please rate National Guard battalions and groups, when deployed, compared to their Active Component counterparts in the role of the Special Operations Task Force (SOTF) or Combined/Joint Special Operations Task Force (JSOTF): (N=237)

Response	Freq.	Percent
About the same as their Active Component equivalent	122	51.84%
Somewhat limited in their capabilities relative to Active Component equivalent	84	35.44%
More capable at some tasks than Active Component equivalent	31	13.08%
No Response	24	

Table C.21
If you are a member of the National Guard, when it comes to operational capabilities, are National Guard enablers about the same as their AC counterparts in these roles when deployed? (N=278)

Response	Freq.	Percent
About the same as their Active Component equivalent	144	64.57%
Somewhat limited in their capabilities relative to Active Component equivalent	35	15.70%
More capable at some tasks than Active Component equivalent	44	19.73%
No Response	38	

Bibliography

19th Special Forces Group headquarters staff, interview with authors, Draper, UT, December 20, 2010.

20th Special Forces Group headquarters staff, interview with authors, Birmingham, AL, December 2, 2010.

46th Special Forces Company Association, home page, January 2010. As of March 1, 2011:
http://www.46thsfca.org/

Baldor, Lolita C., "US Special Forces Show Strain, Says Admiral," *Boston Globe*, February 9, 2011, p. 8.

Blalock, MG Abner C., Alabama Adjutant General, interview with authors, Montgomery, AL, December 2, 2010.

Departments of the Army and the Air Force National Guard Bureau, "Subject: Active Duty Special Work (ADSW) Title 10 Guidance," September 30, 2005.

Government Accountability Office, *Special Operations Forces: Force Structure and Readiness Issues*, NSIAD-94-105, March 1994.

Lippiatt, Thomas F., and J. Michael Polich, *Reserve Component Unit Stability: Effects on Deployability and Training*, Santa Monica, CA: RAND Corporation, MG-954-OSD, 2010.
http://www.rand.org/pubs/monographs/MG954

Morgan, LTC Wayne, *Reserve Component Special Forces Integration and Employment Models for the Operational Continuum*, Carlisle Barracks, PA: Army War College, April 15, 1991.

Newell, Carol E., Paul Rosenfeld, Rorie N. Harris, and Regina L. Hindelang, "Reasons for Nonresponse on U.S. Navy Surveys: A Closer Look," *Military Psychology*, Vol. 16, No. 4, 2004, pp. 265–276.

Office of the Assistant Secretary of Defense for Reserve Affairs, interviews with authors, Arlington, VA, January 25, 2011.

Rhode Island 19th Special Forces Group subordinate units, interviews with authors, Providence, RI, January 11, 2011.

Schemmer, Benjamin F., *The Raid*, New York: Harper & Row, 1976.

Stanton, Shelby, *Vietnam Order of Battle*, Mechanicsburg, PA: Stackpole Books, 2003.

Texas 19th Special Forces Group subordinate units, interview with authors, Austin, TX, January 17–18, 2011.